HAM RADIO

HAM RADIO
A Practical Guide and Handbook

KENNETH ULLYETT

DAVID & CHARLES
NEWTON ABBOT   LONDON
NORTH POMFRET (VT)   VANCOUVER

ISBN 0 7153 7247 5
Library of Congress Catalog Card Number 76–54072

© Kenneth Ullyett and Craft Publications Trust 1976

First published 1976
Second impression 1978

All rights reserved. No part of this publication may be reproduced, stored in a retrieval system, or transmitted, in any form or by any means, electronic, mechanical, photocopying, recording or otherwise, without the prior permission of David & Charles (Publishers) Limited

Printed in Great Britain
by Biddles Ltd Guildford Surrey
for David & Charles (Publishers) Limited
Brunel House   Newton Abbot   Devon

Published in the United States of America
by David & Charles Inc
North Pomfret   Vermont 05053   USA

Published in Canada
by Douglas David & Charles Limited
1875 Welch Street   North Vancouver   BC

CONTENTS

		page
Preface		7
ONE	CQ DX ... CQ DX ...	9
TWO	Ham Activities	26
THREE	CQ-TV: Amateur Television	48
FOUR	Jamboree-on-the-Air	66
FIVE	Getting a Permit	78
SIX	Didah Language	100
SEVEN	Antennas and Channels	110
EIGHT	Rigs and Shacks	121
Appendix of Miscellaneous Data:		135
Technical Terms and Abbreviations		135
Semiconductor Terminology		137
Television Standards		138
United Kingdom Professional Channels		138
Identifying Resistors		139
Identifying Capacitors		141
International Weather Vocabulary		143
RST Code		143
The Amateur's Q Code		144

CONTENTS

Abbreviations for Code-working 145
Ham Prefixes 147
Time Differences 154
International Phonetic Alphabets 155
Useful Addresses 156
International Amateur Radio Union 157
Acknowledgements 159
Index 161

PREFACE

This is the first book on amateur radio to omit circuit diagrams or block layouts. It has no mathematics, no algebra and no formulae, and apart from a number of lists in the Appendix, the text is for general reading. It should give the experienced amateur a resumé of his hobby and, one hopes, introduce the general reader to a pursuit that is absolutely without barriers of language, class, creed or age.

Men and women without any technical qualifications at all can set up a communications receiver, span the globe with their short-wave listening, exchange QSL (message-received) cards, and enjoy worldwide friendships. Then the more experienced amateur can take the necessary technical examinations, together with Morse Code tests (which sensible government legislation makes essential in most countries), and can set up a transmitter, to handle code, phone or perhaps even television or data transmission. American amateurs have their own series of space satellites, which are used for relaying Morse, phone and television. British amateurs have cooperated in bouncing signals off the moon.

There are many amateur-radio emergency services. Explorers have used equipment working on amateur bands, as did Sir Francis Chichester in his round-the-world ocean adventure. There are ham radio clubs throughout the United States and the United Kingdom, in France, Belgium, Holland and Germany, and in India and Japan.

The ether, in fact, is buzzing.

CHAPTER ONE

CQ DX ... CQ DX ...

CQ is the informal amateur-radio abbreviation for 'General call to all stations', and DX is the rather more obvious one for 'Long distance'. 'Working DX' is just one facet of the enthralling hobby that links several millions of men, women and youngsters around the world. These are the radio amateurs – the hams.

Their pastime makes use of the free yet invisible ether, and while some amateurs find the greatest satisfaction in spanning continents and oceans with their home-built rigs, others are concerned with such advanced aspects of ham life as ATV (amateur television with home-built equipment), FAX (facsimile transmission and reception), SSTV (slow-scan TV) and data transmission of a mini-computer nature. To enlarge the horizons of amateur communication, satellites have been circling the earth since 1961. These help to carry even television signals across oceans, and are aptly nicknamed OSCARS – Orbital Satellites Carrying Amateur Radio.

The hobby has been in fashion for over half a century. For example, by the autumn of 1923 more than 100 American amateur calls had been logged in Australia and New Zealand; these were on the 200m band, a channel today within the medium-wave tuning scale of all broadcasting sets, and therefore banned to amateurs. In November 1923 Charles York of Tacoma, Washington, is believed to have set up a link with the American operator of station JUPU in Tokio, Japan, though this record could not be officially confirmed, since it was subject to heavy electrical interference across the Pacific. It is quite certain, however, that in that same month of November the first two-way transatlantic communication took place, between 8AB Leon Deloy in Nice, France, and 1QP-1XAM John L. Reinhartz in the United States.

The 4,650-mile hop from Tacoma to Tokio may have been just a fortunate fluke, but the reliable European-US communication on 27 November 1923 was worked for in a way that exemplifies real ham enthusiasm. Leon Deloy had been captivated since boyhood by Guglielmo Marconi's achievement on 12 December 1901, when he received the now historic Morse 'S' signals (groups of three dots), which were transmitted from his station at Poldhu, Cornwall, to a temporary rig set up in Newfoundland, whose antenna (aerial) was carried by kites and balloons. This transatlantic triumph consolidated the success of the pioneer Marconi Company, registered 4 years earlier; and with the rapid technological advances made in the following 20 years, Deloy believed a transatlantic record could be scooped by an amateur. He devoted his entire energies and a good deal of money, therefore, to 'working the Atlantic'.

He began by attending a convention of the American Radio Relay League (a group of amateurs formed in 1914), where he met John L. Reinhartz; he then bought American-designed short-wave gear, and returned home to Nice with one of the most advanced ham rigs in Europe. He tested his apparatus with the British ham station 2oD in October and then, when the longer dark nights made transatlantic electronic communication less hazardous, he cabled the American Radio Relay League in readiness for the test.

To save time and wristwork in Morse transmission, hams use abbreviations such as the Q code, including QRK ('The readability of your signals is . . .'), and QSA ('What is the strength of my signals?'), and also R ('Received solid'), the aim of every successful ham transmission. A hint of the tension and excitement of this first amateur transatlantic link-up comes from the message the American hams were receiving from Deloy: 'R R QRK UR

CQ DX . . . CQ DX . . .

SIGS QSA VY ONE FOOT FROM PHONES . . . HEARTY CONGRATULATIONS THIS IS FINE DAY . . .' The Atlantic ether had been crossed.

Spanning the Pacific and the Atlantic compelled amateurs (as well as government authorities, owners of cable and wireless networks and those still dreaming of a radio broadcasting system as yet to come into regular operation) to realise that distance was now no object. DX was a fact of life. In a rush of bureaucracy an initial letter was assigned to each country to be used by its hams. The letter was generally the initial letter of the nation itself – A for Australia, C for Canada, F for France, G for Great Britain, I for Italy, and so on. Z was allotted to New Zealand, as N had already gone to the Netherlands (Holland); Cuba was given the phonetic 'Q'; and, for reasons none now can fathom, South Africa was allotted 'O'. Not all these international codings stand today: for example, the Union of South Africa is now ZS, the original U for the United States has given way to W and K, and Argentina has the international prefix LU instead of the R that was originally allotted (because of the phonetic simile of '*Argentina*'.

Britain was among the leaders in this ham pioneering. Station G2OD, which had tested Leon Deloy's transmissions, was situated at Gerrards Cross, Buckinghamshire, and owned and operated by one of the greatest experimenters in the early annals of radio, Ernest John Simmonds. John Clarricoats, OBE (G6CL), secretary of the Radio Society of Great Britain for 33 years, speaks about Simmonds as follows:

On 16th December, 1923, Simmonds made the first two-way contact between England and Canada on 116 metres. On 16th October, 1924, his signals were the first from Europe to be heard in the Antipodes on short waves. A month later on 13th November he became the first Englishman to contact an Australian (3BQ) on short waves. Early in January 1925 he was the first UK amateur to make two-way contact with Mexico when he worked 1B, and in March of that year he was the first to transmit intelligible speech on short waves to New Zealand.

CQ DX . . . CQ DX . . .

Frank Bell of Weihemo, New Zealand, cabled England that he had picked up the signals from G2OD complete with a code-word 'Zinco'. This was on a wavelength of 95m, off the scale of present-day broadcast receivers, and this cable was backed up by one of those fraternal QSL ('Acknowledgement of receipt of signals') cards that are today a busy part of radio-amateur life. The QSL card confirmed receipt of the code-word 'Zinco' and stated: 'Ur the 1st European amateur to hit out 12,000 miles. Ur sigs QSA evy day hr – – – Can hr u evy time u press ur key.'

DX can become a mania. Amateurs annoy their families by sitting up until the small hours, or rising before dawn to get 'sigs thru' thousands of miles. But the G2OD Simmonds contact was typical of a good deal of amateur-radio work in that it broadened our knowledge of the universe. As the *T & R Bulletin* of the Radio Society of Great Britain put it:

> The result of this experiment was far-reaching and outstanding. Radio engineers and physicists were unable to explain this phenomena, as it had always been accepted that short waves were suitable for communication over comparatively short distances. It therefore became necessary to examine these experiments and formulate fresh theories . . . Commercial interest quickly realised the possibilities which would result from the development of low-power short-wave stations for long-distance communication.

Statesmen and brave venturers benefited, too. In May 1925 Simmonds made two-way contact on 23m in daylight (previously all DX links had been during hours of darkness, for sunlight introduced electrical interference) with Charles D. MacLurcan in New South Wales; and using this channel on 4 May 1925, the Prime Minister of Australia sent a message to the British Prime Minister: 'On occasion of this achievement Australia sends greetings.' Two months earlier, G2NM Gerald Marcuse worked SAWJS, the Base station of the Hamilton-Rice expedition on the Rio Branca, Brazil, then exploring the Amazon. The Post Office in London, England, grudgingly gave permission for Marcuse to use high power on a selected number of spot frequencies, as a result of which messages were passed to and fro between the

Amazon explorers and the Royal Geographical Society in London. It was the first of many a headline event linked with amateur radio. Arctic and Antarctic explorers have used equipment operating on amateur bands. While drifting across the Pacific, the Kon Tiki raft was in communication with British and US amateurs. In addition to covering fantastic distances with space-launched OSCAR satellites provided by the American Space Agency, amateurs have bounced signals off meteor trails and from the surface of the moon.

Such things are taken in their stride by a vast international group of hams sending and receiving signals with the fraternal yet scientific aim of improving equipment design. It is a world you can enter, even if you know nothing of radio technicalities, the instant you begin toying with the channel-change buttons and the tuning controls of any good radio receiver having a wide-ranging VHF (very high frequency, or short wave) section.

The everyday 'broadcast transistor' would not generally be suitable for amateur-radio use, because of its unstable and noisy circuitry, even if it covered the appropriate frequency ranges; but with a more ambitious receiver you have only to swing the tuning dial to enter a world as unreal as Swift's Lilliput must have seemed to Gulliver. Lilliputian it is, in the sense of being a world of pygmy wavelengths in the ether, but from the giant's strides these very high frequencies make around the Earth, it is a Brobdingnagian world of gigantic distances.

Naturally, professional engineers and physicists know this world of the radio amateur exists, but they keep it as a thing apart, something to be enjoyed. Big international companies like the RCA, the Marconi Company and many others are peopled by experts in electronics and broadcasting engineering, yet they all have their amateur-radio societies and clubs where in off-duty hours the members can treat short-wave communication as a relaxation. In many countries radio hams are prohibited by law from transmitting regular messages or continuous entertainment, since obviously this would infringe the realms of Posts and Telegraphs, Communications Ministries and broadcasting companies. Of course hams do send messages of a technical nature in the course of experimentation. They transmit music as a criterion for new circuitry and communication systems, and the amateur television hams are busy transmitting pictures – sometimes by satellite. Some hams are chess-lovers, too, but no monitoring station of any government is likely to report an amateur for transmitting such cryptic messages as 'White king's knight blocks his bishop . . . Black moves king to KB2'. Such transmissions are not likely to start a war.

'Amateur' itself is a slightly unfashionable and patronising word, describing one who is fond of something and cultivates it as a pastime, but not seriously or professionally. It comes from Latin, and may remind some of us of schooldays. Then we struggled with *amo* (to love), not realising that one day we should be putting an 'h' in front to create the modern expression 'ham'.

Of course the word 'ham' is derided by serious-minded experimenters on both sides of the Atlantic, particularly by the Radio Society of Great Britain. Esoteric experts feel that to be styled hams is almost as derisory as calling a rare veteran antique automobile an 'old crock'. But there it is. In the USA, alongside the old-established radio amateurs' journal *CQ*, there is the equally official *Ham Radio Magazine*, published in New Hampshire; and this title *Ham Radio* is carried also on magnetic tape for the blind and physically handicapped, by Science for the Blind, in Pennsylvania. *Ham Radio Magazine* goes out to British amateurs through the Radio Society of Great Britain, to European countries from an agency in Sweden, and to the republic of South Africa through Holland Radio of Johannesburg. All this gives the word an official standing of a sort, and to millions of people not connected at all with radio a ham is now not a smoked thigh of hog, nor an over-emphatic actor, but a radio ham – someone who loves the world of amateur radio.

This short-wave world is largely unknown to ordinary TV viewers and radio listeners. To them, moving from one station to another is often simply a matter of button pressing, so they cannot be blamed for not knowing that, with a more sophisticated system, the circuitry can be 'tuned' to a wider range of frequencies than the pre-selected spots of a button switch.

When an electric current is oscillating back and forth in a circuit, it obeys a natural periodicity, rather like the swing of a pendulum but infinitely more precise. Oscillations may vary from the 50 or 60 cycles per second through the filament of a mains lamp to the 10 million cps in the antenna of a television set. This natural electrical resonance is determined by physical reasons, including the length of the wires and other conductors, the length of wire and magnetic effect (inductance) in the coils and the proximity of parts on opposite sides of the circuit (capacitance).

Of course it can be varied – by switching in other circuit elements, or more usually by means of a variable capacitor, or what we used to term a 'variable condenser' – and then the natural periodicity of the circuit is shifted as effectively as when we retune a piano string, or alter the pendulum-length nut in a clock. In the world of radio it means adjusting circuitry so that we can tune from the long waves down to the very short waves; and just what those limits are depends upon the circuitry itself. This is one reason why a simple broadcast receiver is useless for serious ham working.

Tuning is one of the oldest fundamentals of radio, although thousands of hams delicately readjusting their tuning to pick up faint signals probably do not realise they are really doing just what Sir Oliver Lodge did in 1889 when experimenting with high-frequency phenomena. Professor Lodge applied electricity to a Leyden jar capacitor – a glass jar whose walls separated two linings of tinfoil. This primitive 'condenser' stored an electrical charge, and when that charge was released through a wire loop, the resulting oscillations produced a faint spark in a circuit a few yards away. Lodge quickly realised that the greatest effect came when both circuits were of the same size or, as we would say today, 'tuned to the same frequency'. This effect gave Marconi his real breakthrough; his pioneer equipment was devoid of any form of tuning, but was later developed to include what his associates termed a 'syntonic' system. For the first time since Oliver Lodge the transmitter operated at a defined frequency, emitting electromagnetic waves through space, and the receiver picked up the maximum signal when it was syntonically tuned in frequency resonance.

Many years previously, in 1887 and 1888, the German physicist Heinrich Hertz conducted a classic series of experiments with radio waves, on which some radio ham operations and technicalities today depend, although neither Hertz nor Professor Auguste Righi, of the University of Bologna, who extended the Hertz experiments, ever visualised the use of these 'Hertzian waves' for communication. Professor Righi was a neighbour of the Marconi family in Bologna, and the 20-year-old Guglielmo Marconi, after a term or two at the Technical Institute of Leghorn, chanced to read a technical paper by Righi commemorating the work of Hertz, which inspired him to develop a system of wireless telegraphy. Radio hams, who are mostly self-taught in short-wave electronics, will feel a streak of sympathy for the young Guglielmo. As one historian expressed it:

> It was a project for which he was hopelessly ill-qualified . . . He dropped out of full-time education without gaining the qualifications needed for entry to either of the institutions that would have satisfied his father's ambitions for him: the University of Bologna or the Naval Academy at Leghorn. Relations between father and son were further poisoned by the 'scientific' experiments that Guglielmo conducted at home. These were no doubt valuable in developing his experimental skill, but most of them were manifestly trivial, and they simply confirmed the father's conviction that he was harbouring a dilettante . . .

More than 70 years on, there are many young hams experimenting in radio shacks with their home-designed rigs, learning as they go. A good deal of ham work and enjoyment is always dilettante, if you accept the dictionary definition as 'interested in an art or science merely as a pastime and without serious study'.

Within 5 years Marconi had clarified the work of Hertz and Lodge, and developed the matter of frequency selection by

Plate 1 René Klein, founder in July, 1913, of the London Wireless Club which later developed into the Radio Society of Great Britain (RSGB). Mr Klein, then G8NK, is seen at his ham station in 1953, in the very room where 40 years previously the club had been formed

Plate 2 Hiram Percy Maxim, co-founder and long-time first president of the American Radio Relay League, ARRL

Plate 3 Herbert Hoover, Jr, son of the former US President. He served as president of the ARRL from 1962 to 1966

Plate 4 There was a shortwave ham radio station operating at the Vatican, Rome, before the opening of the official Vatican broadcasting station in 1931. The Marchese Marconi is seen at the receiver controls

tuning with his syntonic system. His techniques were incorporated in Patent no 7777 of 1900. This 'four-sevens' patent became the subject of heavy litigation, for, while clearly the origins of the work depended upon Lodge, Marconi's subsequent experiments defined tuning so that it was almost impossible for the great competing cable and electrical companies to avoid infringement. Without tuning there could be no communication beyond a distance of a few yards, and immediately one began tuning, one infringed the four-sevens patent. It is a good thing that radio hams do not have to pay a royalty today to the Marconi Company!

Patent researchers found that Heinrich Hertz had described his spark transmitter as a 'resonator', and Oliver Lodge's own notes of 1889 mentioned a 'syntony experiment'. Nevertheless, High Court patent actions apart, the principle of tuning became firmly established three-quarters of a century ago, and this is now one reason why a radio amateur in Sydney, Australia, can tune in to a ham transmitter in Syracuse, USA, or Surbiton, Surrey. The moment he deviates from the tuned frequency by a mere megahertz, the world link may be broken.

What is he tuning, and what is this measurement of metres or megahertz that runs through all amateur-radio work? Although man now lands on the moon and probes the other planets, regarding voyages beyond the earth's air layer as commonplace, we still do not know very much about the mysterious invisible medium that seems to permeate everything. We call it the ether. When Wordsworth wrote in *Laodamia* in 1817 of 'more pellucid streams, an ampler ether, a diviner air', he did not realise that within a century we should be using this ether as a means of communication. (After the *Titanic* and *Volturno* disasters, the London journal *Punch* of 22 October 1913 published a cartoon of Marconi sitting in a radio cabin, with the following caption: 'Punch to Mr. Marconi: "Many hearts bless you today, sir. The world's debt to you grows fast." ' Now we are getting televised pictures from Mars, all transmitted through this invisible ether!

When one drops a stone into a still pond a ring of waves is

formed; and when the ether is activated by an electronic pulse – it may be the Morse jab of a radio ham or the millionth-of-a-second pulse of a television signal – a ring of waves is also formed. The major difference is that these waves are electromagnetic, and they may go out to 'the ends of the earth', to the moon or to Mars. They are not restricted like the water in the pond, but some of the same physical facts still apply. Float a cork in the pond, drop in the stone, and, although the water itself will not travel out to the cork, the pattern of waves will radiate and cause it to bob up and down. From this pattern we can measure the number of waves passing the cork, and find their 'frequency'. The cork helps us to measure the distance between the peaks (or troughs) of the waves, and this we term the 'wavelength'; it may be a few inches or, as in a storm at sea, many yards. As for the speed of waves in water, we term that the 'velocity of propagation'.

The 'speed' of radio waves is the same as that of light waves, which also travel in this medium of ether; and this an astounding 186,000 miles per second, or approximately 300,000,000m per second. Because the metric system is fundamental in radio, we measure wavelengths in metres and fractions of a metre. Domestic radio users are still accustomed to switching to 'long-wave stations', by which they mean around 1,500m, or to the 'medium-wave band', which extends on the dial or scale from about 200m to 600m. Below that come the 'short waves', and among the many claimants for passage on those short-wave channels are radio hams.

It is obvious even without simple mathematics that if the waves are long, fewer of them will bob that cork every second than if they were short. Thus there is an immediate relation between wave *length* and wave *frequency*. Because of the enormous electromagnetic speed of waves in ether, we need to refer to thousands or even millions of waves per second; and for many reasons it is generally more precise and convenient to speak of the frequency of a transmission rather than its wavelength.

As these waves are created by electronic oscillations in the circuitry linked with the antennas, engineers (hams included) get into the habit of referring equally to transmissions in metres or in frequency. As the waves are transmitted outward from an antenna at 300,000,000m a second, a signal with a tuned frequency of 30,000,000 cycles per second will have waves 10m between crests. It is cumbersome thinking in terms of millions of cycles, so quite early in the development of wireless engineering abbreviations were introduced. We began to speak of kilocycles (thousands of cycles) per second, or even megacycles (millions of cycles) per second. These were written as Kc/s and Mc/s respectively, and hams colloquially referred to the latter as 'Em-cees'.

In the 10m instance mentioned, where there is a distance of 10m between the crests of the waves transmitted at a frequency of 30,000,000 cycles a second, ham jargon would refer either to 'on 10m', or 'on 30 megs'. 'Mega' is the abbreviation for one million, and until the 1960s it would automatically have been inferred that such a transmission was on a frequency of 30Mc/s. Then, to honour the pioneer Heinrich Hertz, the term 'hertz' was adopted instead of 'cycles', although various nations felt that Faraday, Maxwell, Lodge or one of the other pioneers might have been so honoured internationally. 'Kelvin' has been adopted as a unit of temperature based on Absolute Zero (°K), although in fact it was Lord Kelvin who, in a translation of Hertz's technical papers, coined the expression 'aether' waves. The outcome, therefore, immediately the USA followed the German pattern, was that kilocycles and megacycles gave way to kilohertz and megahertz, while 1 hertz (1Hz) became an oscillation of 1 cycle per second. Now 1 kilohertz equals 10^3 cycles per second, 1 megahertz equals 10^6 cycles per cecond, and for the very high frequency (SHF, super high frequency) bands employed for satellite television and radio links we have the relatively new unit, the gigahertz (GHz), equalling 10^9 cycles per second.

The expressions for wavelengths (metres, centimetres and so on) are still used by radio hams, along with Mc/s and MHz, and the various terms are used in that way in this book also, since they are internationally understandable and interchangeable.

CQ DX . . . CQ DX . . .

To avoid interference from (and with) public services such as broadcasting, beacons, and utility communications (including aircraft, fire, police and ambulance services), radio amateurs are allotted bands of frequencies in which to work. In the USA these channels are controlled by the FCC (Federal Communications Commission) in Washington, DC. In the United Kingdom they are regulated by the Home Office, though previously by the General Post Office and the Ministry of Posts and Telecommunications. In most other countries the bands are controlled by the appropriate Ministry or Department of Posts and Telecommunications, since the false notion dies hard that ham channels could short-circuit the official and sometimes profitable channels on which messages and radio and TV programmes are handled.

Bands allotted to amateur transmitters are in the following ranges (MHz):

1·8–2·0	430–432 and 430–440
3·5–3·8	1,215–1,325
7–7·10	2,300–2,450
14–14·35	3,400–3,475
21–21·45	5,650–5,850
28–29·7	10,000–10,500
70·025–70·7	24,000–24,050
144–145 and 144–146	

Confusing to newcomers to ham bands, these figures do not mean that after obtaining a transmitting licence one can simply plug in a microphone or a Morse key and blurt out 'Hello CQ' on all channels. There are many different types of emission, which will be examined in detail in Chapter Two, and not all may be used on all bands. There is also a limit to the maximum transmitter power, for what may be reasonable on one frequency channel could possibly result in interference on another. It may surprise newcomers to the hobby to learn how small this power is, whether measured at the input to the transmitter or as the radio-frequency output peak waveform envelope power.

The average single bar of an electric radiator takes 1,000 watts

(1kW) of energy, and electric light bulbs range from 25 to 200W. In the United Kingdom a ham transmitter operating on the 1·8–2·0MHz band is restricted to *only 10W input*, and even in the 14–28MHz bands there is a power limit of 150W, which can give an RF (radio-frequency) peak output of 400W. It is a near-incredible truism of amateur working that with such small powers one can span continents. These band limits are set for amateur transmitters, but of course there is no restriction on ham receivers, and while the operators of stations are very restricted in what use they may make of any information received, there are no bands on which they may not listen.

From time to time the Home Office varies the allocations, and as we shall see in Chapter Five there have been relaxations so that it is now possible to obtain a type of amateur (sound-only) licence for all bands from 144MHz upwards without passing a Morse Code test. Knowledge of Morse has been a sort of Achilles heel to some potential amateurs, who are fascinated by electronics but not by the dots and dashes of Samuel Finley Breese Morse.

The Radio Society of Great Britain has continually made strong representations on behalf of amateurs in the United Kingdom to get a fair share of frequencies; and in the United States the ARRL (American Radio Relay League) is in almost continual conference with the FCC regarding ham bands, and the position is always rather complex. According to the ARRL definition, a radio amateur is 'a duly authorised person interested in radio technique solely with a personal aim and without pecuniary interest', and, as defined in the ARRL's own radio-amateurs' handbook, there are currently five available classes of amateur licence available to permanent US residents. These are Novice, Technician, General ('Conditional' if taken by mail), Advanced, and Amateur Extra Class. Extra Class licences have exclusive use of the frequencies 3·5–3·525, 3·775–3·8, 7·0–7·025, 14·0–14·025, 21·0–21·025 and 21·25–21·270MHz. Advanced and Extra licencees have exclusive use of 3·8–3·89, 7·15–7·225, 14·2–14·275, 21·27–21·35 and 50–50·1MHz.

Major US ham bands are around 10m (28–29·7MHz), 15m,

20m (the 14MHz band), 40m, 80m and 160m. Maximum amateur input power is limited generally to 1,000W, and in certain cases for novices to 75W. Just as in the United Kingdom, where certain spot aviation frequencies, including 144, 144·09, 144·9MHz and others, must be avoided, so there are US restrictions on bands shared with the Government Radio Positioning Service, which naturally has priority; and hams in Guam, Palmyra, American Samoa and Wake Island are prohibited from operating on frequencies from 3·9–4MHz and 7·1–7·3MHz. There is a changing and lengthy list of limitations in the 160m band (1,800–2,000kHz), and the ARRL issues charts of such restrictions. No American ham should go on the air without first studying the ARRL *License Manual* ($1.00 postpaid in the US, and available from the RSGB and the publishers of the *Short Wave Magazine*, 55 Victoria Street, London, SW1, for amateurs outside the United States). British amateurs may obtain similar vital information from the RSGB, 35 Doughty Street, London, WC1N 2AE. A list of all ARRL publications regarding bands can be obtained direct from the American Radio Relay League, Newington, Connecticut, 06111.

There are concessions to US hams visiting or working in many countries, including the Argentine, Bolivia, Brazil, Canada, Guyana, Israel, Kuwait, Panama and elsewhere, and the ARRL supplies the latest information to members.

The amateur licensing situation in Canada naturally runs a close parallel with that in the US, but there are essential differences. The following is an abridged account of the basic ARRL review, and copies of the latest regulations may be obtained for a nominal fee from the Department of Communications, Ottawa. There is a useful book, *The Canadian Amateur Radio Regulations*, issued at $2.55 from CARF, Box 356, Kingston K7L 4W2, Ontario, Canada.

The ARRL informs neighbouring US hams (who of course might be affected by any irregularity in a Canadian amateur's transmissions) that amateurs are now not restricted as to age, and may take an examination for an Amateur Radio Operator Certificate at one of the many regional offices of the Department of Communications. There is a Morse Code test of ten words per minute, and oral and written examinations on basic transmitter theory. The fee currently is $10.00 annually, and this entitles the applicant to transmit continuous-wave code (see p 45) on all authorised Canadian channels, and to transmit phone (telephony) on those approved bands above 50MHz.

A useful introduction for French-speaking Canadians may be obtained from *Comment Devenir Amateur*, published by VE2BTG Guy Cadieux, 924 20th Avenue S., Ville de St Antonio, PQ, Canada, at $2.50.

Six months after taking the first examination, and if the station has been regularly operating on continuous wave on frequencies below 29·7MHz, the operator may apply to have his permit endorsed for phone operation in that lower band, which at present is 28–29·7MHz. There is a study book, *Ham Handbook for Advanced*, issued at $5.30 by ARTA Publishing Co, PO Box 571, Don Mills, Ontario, and this type of technical guide (like the ARRL and RSGB handbooks) is almost essential for amateurs wanting to take subsequent examinations for the Advanced Amateur Radio Operators' certificates. An indication of the technical standard required in examinations held in the North American continent is given in Chapter Five. As in the United States, the maximum input power to the final output stage of a Canadian ham transmitter is limited to 1,000W, which of course is far and away greater than that applicable in the tighter confines of the UK and European ether generally. But, as every experienced amateur knows, high power is not always necessary to achieve DX.

CHAPTER TWO

HAM ACTIVITIES

Precisely what the new ham finds enthralling in his hobby depends upon his own nature and mental make-up. Some may want to provide an essential service, as did G2NM Gerald Marcuse back in 1925, when he supplied the only link between the Royal Geographical Society in London and the Hamilton-Rice team exploring the Amazon. Amateurs also provided emergency links when an earthquake hit Alaska (1964) and the tropical storm 'Agnes' brought the disastrous Dakota floods (1972).

Other amateurs concentrate their energies on electronic technology, on occasions forestalling and outrivalling the professional engineers. Over 50 years ago radio-transmitting amateurs were the first to use crystal-controlled oscillators for frequency stability, proving systems that were adapted for present-day broadcasting of radio and TV. They were also the first to develop a means of neutralising power amplifiers in transmitters, which reduces parasitic oscillation, and were consequently responsible for the radio industry introducing tubes (valves) such as the frame-grid triode PC97 and the PC95/6ER5 beam triodes, to help broadcasting engineers use the most stable master-oscillator power amplifiers. Youngsters who build their first basic receivers at school soon learn about what US engineers term 'regeneration' and British students know as 'reaction'. Put simply, it is the technique in which part of an output signal while at HF can be fed back to the input. Major Edwin H. Armstrong, while still an amateur, developed his understanding of regeneration in radio circuits.

There is always a great sense of adventure in these 'firsts'. Dr Lee de Forest – known as the Father of Radio to millions of Americans because, before the transistor era, he developed a tube with a grid (and which could therefore amplify and oscillate), after the basic two-electrode thermionic tube had been invented and patented by Sir Ambrose Fleming in 1904 – was fond of telling the following story about Armstrong:

> During the fall of 1913, I presented a paper on The Audion Amplifier [he referred to his tube as an Audion then] before a meeting of the Institute of Radio Engineers at Columbia University ... My demonstration of the crashing sounds emitted from my loudspeaker when I dropped a handkerchief on the table before the telephone receiver serving as my pick-up microphone, aroused great astonishment and applause. On that occasion young Edwin H. Armstrong, wrapped in deepest mystery, had a small, carefully-concealed box in an adjoining room into which neither I nor my assistant Logwood were permitted to peek. But when he, Armstrong, led two wires to my amplifier input to demonstrate the squeals and whistles and signals he was receiving from some radiotelegraph transmitter down the Bay, I thought we had a pretty fair idea of what the young inventor had concealed in his box of mystery.

Later in the lives of both men, in January 1921, historic litigation began in the Federal courts as the Westinghouse Company, licencees under the Armstrong regenerative-circuit patent for feedback, sued the Radio Telephone & Telegraph Company for infringement. The battle between the Radio Corporation of America, AT & T and De Forest Radio Company, and the respondents' Radio Engineering Laboratories Inc, went on for 19 years and became a *cause célèbre* among the annals of patent jusrisprudence.

In more recent times radio amateurs have led in the development of high-gain beam directional aerial systems, and pioneered a space-saving single-sideband system used professionally in communications and broadcasting. Hams carried out sustained scientific observation throughout the International Geophysical Year (1957-8) and during the International Quiet Sun Years (1958-9), and have participated in space research investigation since the first Sputnik was launched by the USSR in 1957. The 1971 World Administrative Radio Conference for Space Telecommunications recognised the value of amateur work by officially defining the amateur satellite service.

When K2UN came on the air in the summer of 1948, it became obvious that the ham movement could be an influence for world peace and understanding. K2UN, call-sign of the United Nations Amateur Radio Club station, was chosen as signifying 'Come to the United Nations'. The station, at Lake Success, opened with two 1,000W transmitters: one TX (transmitter), working on the 40m and 80m bands, fed a doublet antenna to cover a wide area of the United States; and the other had a rotary beam array and was planned for the DX 10m and 20m bands, to reach out directionally from Lake Success to the rest of the ham world. A very sophisticated operating console was provided, with an instrument for showing the amount of local interference, so that this information could be passed to other amateurs wanting to work K2UN.

This TX was intended to be symbolic of the United Nations Amateur Radio Club's aims, which were 'to preserve and foster the spirit of fellowship among the radio amateurs of the world: to promote international interest in the UN's role of building a better world, and to build prestige for the United Nations through friendly cooperation with radio amateurs everywhere'. It was not intended that propaganda should be transmitted on K2UN, but that station operators among the fifty original secretariat members of the club should answer on-air any individual questions about UN work.

Operating the station on the opening day in 1948 was W2KH George W. Bailey, then President of the International Amateur Radio Union and executive secretary of the American Institute of Radio Engineers. His function as the senior member of the IARU emphasised perhaps for the first time to many thousands

HAM ACTIVITIES

of hams throughout the world the work that has been done by the Union since it was formed in 1925.

Outstanding support for the ham movement is given in the United States by the ARRL, and in the UK by the RSGB, but organisation is in fact worldwide. The International Telecommunication Union (ITU) is a specialised agency of the United Nations, and for technical and administrative reasons its work is split into three world regions: Region 1 comprises Europe, Africa, the whole of the USSR and certain parts of Asia; Region 2 North and South America; and Region 3 the rest of the world, including Australia, New Zealand and Japan.

Under the general umbrella of this organisation come the national amateur-radio societies, some of which, like the ARRL, have large memberships and histories going back to the days before World War I, while others are new. More than eighty national amateur-radio societies are members of the IARU (International Amateur Radio Union), which has its headquarters in Geneva, Switzerland; its two main functions are promoting and coordinating two-way radio communication between the amateurs of the world, and representing their interests at the ITU conferences when such important matters as ham bands are decided.

The IARU was represented by observers at the ITU conferences held in Madrid (1932), Cairo (1938), Atlantic City (1947), and Geneva (1959, 1963 and 1971). Following the Atlantic City conference, which was in session for 6 months in 1947, the European national amateur-radio societies formed themselves into a distinct division of the IARU to represent the special interests of the hams of ITU Region 1. This IARU Region 1 Division, as the organisation is called, consists of forty national member societies, all of which contribute at an agreed rate per licensed member towards running costs. The *IARU News* is published regularly, and distributed to all member societies. The successful formation of the IARU Region 1 Division was followed by similar moves in Regions 2 and 3, and each now has its own organisation of national societies in its area.

HAM ACTIVITIES

Headquarters of Regions 2 and 3 are situated in Lima, Peru, and Melbourne, Australia, respectively. Currently the Vice-Chairman of Region 1 is André Jacob (F3FA), in Pavillons-Sous-Bois, France; the Secretary is Roy F. Stevens (G2BVN), 51 Pettits Lane, Romford, RM1 4HJ, England; and the executive includes such experienced amateurs as Axel Tigerstedt (OH5NW) Finland, W. Nietyksza (SP5FM) Poland, H. Walcott-Benjamin (EL2BA) Liberia, Janez Znidarsic (YU3AA) Yugoslavia, and Lt-Col Per-Anders Kinman (SM5ZD) Sweden.

There is a headquarters amateur-radio station, 4U1ITU, on the third floor of the International Telecommunication Union building in Geneva, and in 1974 this was re-equipped by the ARRL and RSGB to celebrate the 6th World Telecommunication Day. A list of member societies of the IARU is given in the Appendix (p 157). Additionally, there are 'VHF Managers' – experienced amateurs appointed in each country to help supervise good working on VHF bands. In Great Britain the RSGB-appointed VHF manager is G. M. C. Stone (G3FZL), 11 Liphook Crescent, London, SE23. In Eire the manager is A. Latham (EI6AS), 92 Glenareary Estate, Dun Laoghaire, Co Dublin; in France, Jacques Talayrach (F9QW), 86 rue de Vieux-Chateaux, 91 Yerres; in Italy, F. Armenghi (I4LCK), Via Sigonio 2, 40137 Bologna; and in Germany, H. J. Schilling (DJ1XK), 775 Konstanz 16, 1M Gruen 7. The headquarters of the IARU itself is at 2 rue de Varembe, Geneva, the postal address being Box 6, 1211 Geneva 20, Switzerland.

The International Union aids projects, including amateurs' satellites in the OSCAR series, and it approves such projects as the Radio Amateur Satellite Corporation (AMSAT), a non-profitmaking unit founded in the United States in the late 1960s. The amateur satellite programmes are supported entirely from donations and membership dues. There is a quarterly *AMSAT Newsletter*. Amateurs may obtain details from the Membership Committee, AMSAT, PO Box 27, Washington DC, 20044 USA.

It is wonderful how ham activities continue despite wars and civil and political strife. Amateur-radio human interest cuts right through terror and tyranny. For example, while the People's Republic of Bangladesh was going through agonies of formation, starvation and flood, a central nucleus of technicians continued to operate amateur-radio stations, some of which were vital links in flood-disaster communication chains. Bangladesh was able to reorganise, and to join the ITU in September 1973. In much the same way the Cyprus Amateur Radio Society was formed in 1961 to stimulate amateur activity and set up a QSL-card bureau to handle incoming cards, as well as to represent Cyprus amateurs within Region 1 of the IARU, of which of course the society is a member. All this ham activity continued despite the civil, political and military problems Cyprus faced in the 1970s. While Nicosia, Limassol, Paphos and Famagusta were being headlined in the world's press, and as Europe feared for the safety of citizens in Cyprus, the radio amateurs were able to play their disciplined and properly organised part in restoring normal life in the island. The society's president lives in Famagusta, and representatives include 5B4AC in Nicosia, 5B4AH in Famagusta, and 5B4AV in Limassol. Immediately the military situation permitted, this brave section of Region 1 was able to resume the issue of its ham-operating trophy, the Cyprus Award. The Awards Manager can be contacted through PO Box 1267, Limassol.

A grave problem was faced at one period of postwar history by the radio amateurs of Czechoslovakia, but the CRCC (Amateur Radio Club) was able in 1973-4 to celebrate both the fiftieth year of public broadcasting and the fiftieth anniversary of a first attempt to organise radio amateurs in the country. Displays were held in Prague and Bratislava, and some 150,000 visitors came to see the Club station OK5OR in operation.

The SARL is an active member society of the IARU, with its HQ in Cape Town, Republic of South Africa; and in the same way the RSR in Salisbury, Rhodesia, is a member society. Differing political creeds have absolutely no place in world amateur radio – indeed, the fraternity of short-wave communications helps promote international harmony among real people.

The administration of South Africa has relaxed some restrictions regarding amateur working, and the IARU has recorded the fact that the administration now issues licences to anyone who, except for not passing a Morse Code test, would have qualified for the issue of a full amateur licence. Licences are now issued in the ZR series, to distinguish them from the full ZS licence. The ZR licensee may use radio telephony only on frequencies of 144MHz and above, and there are frequency limits at present that seem to preclude South African amateurs from communication via OSCAR satellites.

Orbital predications of the OSCAR series are given from time to time in *Region 1 News*, the journal of that division of the IARU, the information showing orbit number, Equator-crossing time and Equator-crossing latitude. For AMSAT-OSCAR 6 (see Fig 1), an orbital predictions book is also published by *Ham Radio Magazine*, of Greenville, New Hampshire 03048, for approximately $3.00, plus postage from the USA. This contains equatorial crossing times and longitudes for all the 4,183 passes of OSCAR 6 during the year, with an accuracy estimated at better than 10 seconds.

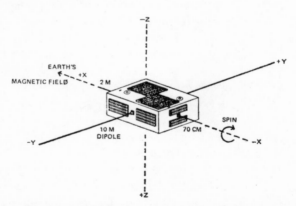

Fig 1 Configuration of the amateur-radio satellite OSCAR 6 in orbit. The unit weighs 40lb (18·1kg). The dipole aerial is of ½in carpenter's measuring tape, so that it could be folded alongside the space craft during launch

While there is no mystery about IARU addresses, amateurs sending QSL cards and technical queries to hams in the USSR must address them 'Box 88, Moscow D-362, USSR'. There are some 46,000 licensed amateurs operating in the Soviet Union, and 4,500 clubs, with more than 17,000 members. The only way to reach them, for foreigners, is via 'Box 88, Moscow'. Power ratings allowed for the three classes of licensed transmitting amateurs are set at 200W (Class 1), 40W (Class 2), and 10W (Class 3).

Few knew what happened inside 'Box 88' until a distinguished New Zealand amateur, J. L. Carrell, ZL1HL, went there in 1973 and reported back to the NZART and IARU as follows:

> My journey to Box 88 was by Metro and taxi . . . An interpreter who works at the nearby Aeronautics Club was on hand for my visit . . . I believe the club, like sports groups in the USSR and other East European countries, has substantial financial support from the Government. At the club headquarters a full-time staff of eighteen is employed in a building of about 1,800m^2 (some 19,000sq ft), on three levels. Apart from the administrative offices the building houses a library of 48,000 reference books (not including magazines) and some 12,000 technical papers, a reading room, a lecture theatre the size of a small cinema, a lecture room, laboratory and workshop, and of course the QSL bureau. This bureau has a full-time staff of four women who handle annually some 2½-million cards, including about 8,000 to and from New Zealand. While visiting the bureau I picked out cards from Soviet amateurs destined for Auckland hams I know, and was able to deliver them in person on my return home . . . A club headquarters station is situated 35km (about 21 miles) away, and has a 1-kW transmitter capable of operating on each of the five HF bands, and a beacon on 144·5MHz . . . The central club organises contests, issues contest award certificates and publishes a monthly magazine *Radio*, which has a circulation of one million copies. The contest to follow my visit was for operation on 144MHz and 342MHz, there being sixteen stations in the contest, with four operators per station.

It must be stressed as an example of international ham fraternity that the very first direct satellite link between the Soviet

Union and the United States was accomplished not by a huge commercial satellite such as we are now accustomed to for trans-ocean television and phone, but by the amateur satellite OSCAR 4.

In emphasising the lure of amateur radio, the ARRL says:

> The reason that amateur radio is often called the most satisfying and thrilling of all hobbies is that it offers something for everyone. It is 'all things to all men'. . . . For example, you may be a tinkerer – you may like to play around with gadgets, build them up, make them work. Amateur radio is the ideal hobby for the tinkerer who likes to go into the 'why' of the things he builds. You may be a 'rag-chewer'. Most hams are. The most enjoyment you know may come from getting together with a crowd of good fellows and talking over everything under the sun. Amateur radio is full of confirmed addicts of the conversational art; indeed, there is even a Rag-chewers' Club, with a membership certificate signed by 'The Old Sock' himself for those who qualify.

This is correct from the ARRL standpoint, but in the United Kingdom and some other countries rag-chewing is not encouraged, except on a limited basis for conveying technical information specifically linked with the ham experiment in hand.

Enthusiasm in the amateur is dealt with by the ARRL in outlining activities that test the mettle of each type of ham. The Communications Department of the ARRL has set up a comprehensive organisation complete with trunk lines and field officials, which is paralleled by that of the Radio Society of Great Britain, enabling amateurs to compete to see who can relay the most messages. 'DXing is actually a glorified form of fishing' is how the ARRL puts it. 'It takes endless patience and skill, but to the true fisherman it has a zest nothing else in the world can equal – and it is a sport you can indulge in any day, any season of the year.'

There are Sweepstakes for amateurs to take part in throughout Canada and the USA, while mobile days are organised in much the same way by amateurs grouped in the RSGB. In the USA also there is the Military Affiliate Radio System (MARS), operated jointly by the air, naval and military branches of the

Plate 5 Memorial at Greenwich, Connecticut, to station 1BCG, built and operated by members of the Radio Club of America, the first in 1921 to transmit on short waves across the Atlantic. *Left to right*: Major Edwin H. Armstrong, George E. Burghard, Paul F. Godley and Ernest V. Amy, all of whom took part in the pioneer transmission

Plate 6 Typical British ham shack in the early 1920s before AC mains served nearly every area of Britain. This rig, with ebonite-panel transmitter and glass-panel instruments and receiver, was powered by storage batteries (right)

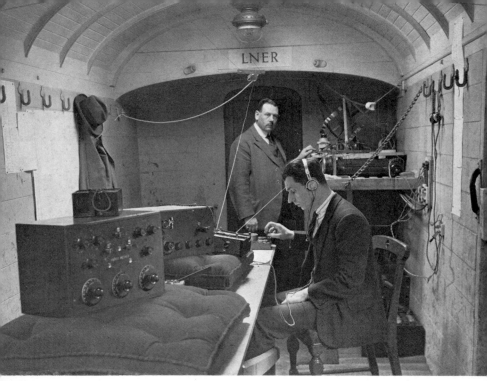

Plate 7 Ham radio coach of the LNER Scottish Express in July 1924, when communication was effected all the way along the route from Bedford to Newcastle-on-Tyne on 182.5m. At the key is Maurice Child, while Leslie McMichael (also a pioneer British amateur) adjusts the transmitter

Plate 8 The Lord Mayor of Birmingham, England, in 1949 opening by radio the first amateur convention to be held in the Midlands after the ending of World War II

armed forces, which gives hams the opportunity of assisting the defence forces and at the same time acquiring training in military communications. The ARRL says:

> The communicating experience an amateur acquires, and the organized networks in which many hams participate, become in time of disaster of untold value to the community and nation. Let a hurricane or an earthquake or a flood destroy normal lines of communication, and hundreds of amateurs are ready to step in and provide emergency circuits for the Red Cross, civil defence, military and municipal agencies. So important, in fact, is the amateur's work in emergency communications that the FCC has set up special rules for a Radio Amateur Civil Emergency Service, in which public-spirited amateurs can enroll as a part of civil defense efforts. These local networks engage in practice drills regularly, hoping that disaster will never strike their community, but determined to be fully prepared.

When he was Governor of New York State, Nelson Rockefeller paid tribute to hams via the press, radio, and TV:

> The amateur radio operators perform a most valued service, for which all New Yorkers are grateful. Never was their value more clearly demonstrated than during the Southern Tier flooding of June 1972, one of the worst natural disasters ever to strike our State. Without regard for personal safety in many cases, amateur radio operators gave important support to official government agencies and the military in rescuing and protecting hundreds of thousands of flood victims in the stricken areas.

Ham radio has also been of vital assistance in many lesser cases than a New York State flood. Here are some recent examples from *QST* records.

A call for rare drugs for a 12-year-old child sent by CN8BF (Morocco) was received by VE2BRW (Province of Quebec) at 1831 GMT on 20m. Since conditions were poor, 6W8DY (Senegal Republic) helped relay the information. With the help of the Red Cross and a local doctor, VE2BRW located the drug in Montreal and arranged for its shipment via Paris, France, the following day. Without the drug the child would have lived only 2 weeks.

HAM ACTIVITIES

While travelling from Kansas City to Bonner Springs, WØQJU spotted five cars that had skidded off the road at different locations, and a four-car accident blocking the turnpike. All were called in through KØBXF to turnpike officials.

At 1330 K3ICH/mobile was hit broadside by a speeding automobile, which injured his wife. His call for assistance on the WB4QFP repeater was answered by K4CGY, who notified police and ambulance. Assistance arrived within 3 minutes.

A fire at a home in Winnipeg seriously injured the owner's son and a housekeeper while the parents were en route to a vacation in Mexico. Attempts were made to contact the couple by amateur radio; and with the help of Mexican and Canadian officials, and hotel and radio-station personnel, they were located in Acapulco. Several Canadian, Mexican, US amateurs and the YL-ISSB net took part in this operation.

WA8BRD/mobile (Ohio) arrived at the scene of a hit-and-run accident moments after it had occurred. The victims' car was on fire and they were trapped in it. Through the WB8CQR repeater (Michigan), he called W8GRG, who relayed the report to the Cleveland Fire Department, which arrived on the scene within 3 minutes. In the meantime WA8BRD brought the fire under control with extinguishers, and then, with the help of others on the scene, freed the occupants. Aside from a bad scare, they were uninjured.

On hearing of a plane crash into an Almeda apartment house, WA6AGA (California) and WB6s GWQ and RPK immediately conferred, and then committed the resources of the Grizzly Peak VHF transmitter for emergency communications with the Red Cross. W6NKF (California) and K6KAP went to the scene with portables linking them to WA6GCS at the Almeda Red Cross headquarters. Other amateurs reported to the scene throughout the night as the fire blazed, while still more made themselves available for the work of cleaning up and the search for victims, and for providing the radio links needed by the Red Cross. This operation took three days.

These are just six examples out of thousands, which can be

HAM ACTIVITIES

matched in most countries where hams have organised what US amateurs know as ARPS, Amateur Radio Public Service.

Other amateurs achieve a different sense of achievement in technical developments. In the United Kingdom some of the most erudite amateurs – naturally, one might suppose – are attached to the big electronic groups. These include the Mullard Group, Mullard House, Torrington Place, London WC1; the Marconi Company, Marconi House, Chelmsford, Essex; the Racal Group (Services), 21 Market Place, Wokingham, Berks RG11 1AJ; EMI Ltd, 135 Blyth Road, Hayes, Middlesex, UB3 1BP; the Plessey Company Ltd (club correspondence c/o Parker PR Associates Ltd, 22 Red Lion Street, London WC1R 4PX); the Post Office, 23 Howland Street, London W1P 6HQ; and the British Broadcasting Corporation, Portland Place, London W1A 1AA.

As mentioned in Chapter One, there is a large amateur-radio club attached to the RCA, and it was the RCA, of course, which built America's operational weather satellites. Members of the RCA amateur-radio club are in a specially favourable technical position, and an example was set by the veteran radio amateur Wendell G. Anderson, who in his professional capacity is staff technical advisor for RCA Government and Commercial Systems in Moorestown, NJ. He built New Jersey's first amateur equipment to receive TV pictures of the earth and its weather from the satellites now orbiting hundreds of miles overhead. This receiving station was put together primarily from a 30-year-old ham radio set plus the sort of equipment accumulated by most amateurs. The total outlay was about $200 (well under £100 at that time). The antenna was fashioned from a piece of wire mesh and a 30ft length of copper tubing held in place by wooden dowels and fastened to a clothes pole in the garden. Secondhand electric motors costing $10 each help rotate the antenna to keep in line-of-sight of weather satellites. Wendell Anderson has corresponded with more than 200 other hams around the world, as far away as Canada, India, Italy, the Netherlands, West Germany, Turkey and South Africa.

HAM ACTIVITIES

The key to amateurs being able to pick up the satellite TV picture is that the RCA-built spacecraft can transmit images of local-area weather directly to relatively simple ground equipment. Professional weather-recording receivers are employed in some fifty countries in every continent, and now short-wave hams use their ingenuity participating in this branch of space work, in a personally exciting way.

Another enthusiast of the same electronics group is A. W. (Tony) Slapkowski, WB2MTU (see Fig 2), who in his everyday work is concerned with space electronics of the Apollo 11 variety. In the 1960s he discovered that many amateurs were interested in US space operations, yet had only limited access to information. He originated a 'net' of ham information, and this reached a climax in July 1974, when the Governor of New Jersey, Brendan T. Byrne, proclaimed an Amateur Radio Week in honour of the fifth anniversy of Man's first landing and walk on the moon (July 1969).

The FCC authorised the use of the call-letters WM2OON to be used on all amateur-radio frequencies worldwide during

Fig 2 RCA Radio Club member Tony Slapkowski sends these QSL cards to amateurs contacting him on space weather satellites and similar matters

Fig 3 QSL card of G2BVN, Secretary of Region 1, International Amateur Radio Union

this commemoration, and calls were handled by eleven ham stations centred on Tony Slapkowski's 5 acres of garden and woodland in Sicklerville, NJ. QSL cards with a design commemorating the Apollo 11 mission were issued as an acknowledgement of hams working with WM2OON. Over 14,000 ham operators throughout the world participated in this special moonshot amateur-radio net, and Slapkowski's own WB2MTU remains a useful centre of information for amateurs wanting to keep up with short-wave and space-communications information.

Amateur television and techniques of slow-scan TV are described in Chapter Three. FAX (facsimile) is a branch of the hobby that is progressing in the United Kingdom, inspired no doubt by the US pattern of permitting this type of transmission in rather restricted 2m and 6m bands. Facsimile is really a reversion to the very earliest type of document transmission such as the Fultograph, an invention of Captain Fulton used experimentally by the BBC in the early 1930s.

Material to be transmitted is clipped to a slowly rotating drum,

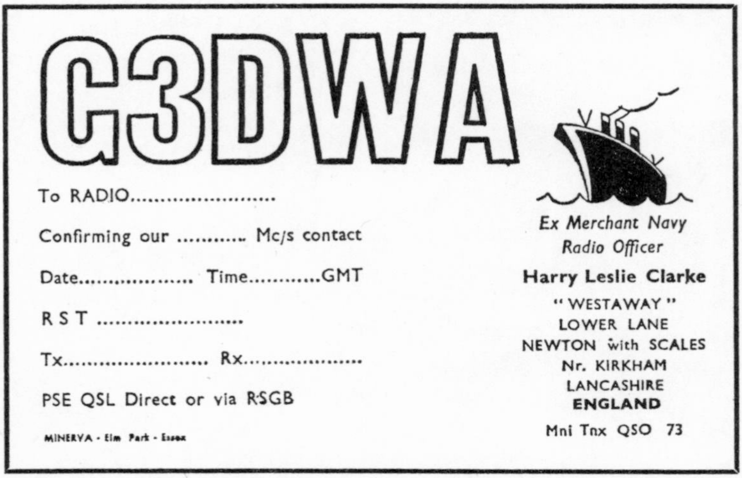

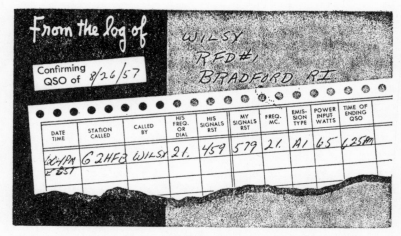

Fig 4 QSL card of RSGB member, an ex-Merchant Navy Radio Officer

Fig 5 Part of the official ARRL log is included in this QSL card of a Bradford, RI, amateur

scanned by a very narrow light beam. A photo-electric cell acts on the light reflected from the material surface and transforms the myriad of variations into a modulated direct current, the DC then running a voltage-controlled audio oscillator. One single drum revolution gives a one-line scan, and of course, as the drum revolves, it is moved laterally or moves the pick-up system. The information is broadcast as an audio tone, and the receiving end is detected and amplified, then applied to a receiver drum moving in synchronism with that at the transmitter. The picture or other information is 'drawn out' in a number of ways. The DC voltage may be used to vary a light beam directed at the drum, or cause a voltage-sensitive paper to change colour electrostatically at the pen-point. This was the Fulton system. Some hams build very sophisticated FAX apparatus in which the information is transmitted and printed out flat, without a drum. It is normal to expect a picture speed of 2–3 minutes; with slow drum speeds and optimum use of the available bandwidth, FAX hams can get pictures of photographic quality. Amateurs who take greater delight in construction than in reception convert commercial Telefax transceivers to this type of amateur-radio work.

In all forms of ham operation one comes upon initials, including AM (amplitude modulation), FM (frequency modulation), CW (continuous wave) and SSB (single sideband), and these are to be found in nearly every branch of amateur activity, from Morse transmission to home-built TV. Those who consult books on radio theory will already be familiar with these terms. A potted explanation will help the general reader to follow some of the descriptions in following sections of this book.

Driving along with the car radio playing, you are overtaken by a motorcyclist whose machine has not had the ignition circuit electronically suppressed. You hear a loud crackling in your car-radio speaker, which fades as the machine moves more than 10yd or so away. This is 'spark' transmission (strictly speaking, illegal under the motoring regulations of many countries), the type of radio used for communication at sea in the very earliest days. A spark-gap, or, later, the carbons of an arc, was fed with a voltage supply, and the resulting electromagnetic disturbance in the ether could be detected many miles away – indeed, across the Atlantic. The harsh crackling note of the spark was intermittently started and stopped by the Morse key, resulting in code dots and dashes. All that is past history. Today there are very many thousands of sets using the ether all at the same time – civil and military communications networks, broadcasting apparatus, beacons, weather stations, walkie-talkies and hams. There simply would not be space for such an all-blanking and inefficient means of creating a signal. Moreover, its use is limited to being chopped up into short or longer bursts for Morse Code dots and dashes. 'Spark' cannot have anything such as speech or video frequencies imposed upon it, to modulate it.

Instead we use tubes (valves) or even transistors for relatively low-power transmission. Hams make use of large tubes for powers of around 1,000W, while the huge tubes used by commercial stations handle so much power that their anodes may need air- or water-cooling systems.

Whether large or small, they are coupled to circuits tuned to resonance electrically, which instantly allow the build-up of oscillations. This circuitry is coupled to the antenna and ground (earth) section of the transmitter, so that the resulting oscillations (or perhaps harmonics of them) are conveyed to the ether. Unlike spark transmissions, which have the audio element of the spark note built in, they are at very high frequencies, far above audio limits, as was mentioned in Chapter One. The audio limit of the human ear is usually around 15,000 cycles per second (15Kc/s), while with some people it is as low as 5,000c/s. However, the train of waves we send out from the antenna is at millions a second.

If we could see the pattern of waves (which, mercifully, no human eye can do), it would appear as a continuous stream of sine waves. Hence, in ham jargon, the expression 'continuous-wave' and the abbreviation CW. The electromagnetic kick the oscillator and radiator (antenna-earth system) give the ether, produces a steady continuous stream of waves rather like a comb with rounded teeth. The trace is that of sine waves, and from the terrors of trigonometry it will be recalled that a sine wave is one exhibiting simple harmonic motion.

The simplest way to use it for communication is to chop the wave up into dots and dashes, to produce radio-telegraph signals of the form classified in amateur-radio regulations as 'A1'. Then, at the receiving end, this interrupted stream of CW is detected and mingled (heterodyned) with oscillations introduced into the receiver. These local oscillations are slightly off-beat with respect to the received frequency, so the resultant is within audio limits. The output note or tone can be varied by regulating the frequency of the 'beat-frequency oscillator' (BFO) to get a pleasant code tone in the earphones.

Unfortunately even chopping the stream of CW has its problems. With systems such as 'A2' the CW stream is modulated or shaped at the outset by the waveform of an audio oscillator, so that in reality the continuous stream of sine waves is given an envelope of audio frequency. This is known as amplitude-modulated tone telegraphy. We call it amplitude-modulated, because it is indeed the amplitude (height) of the sine waves that is modulated by the tone envelope, not their frequency. It might be thought that if a transmitter is oscillating at, say, 3,600kHz, the output energy would instantly be at 3,600kHz each time the Morse key is pressed down. Unfortunately, even an electrical circuit has a time delay, minute though it may be, so that each time the circuit is switched on, a small amount of energy is actually expended as the transmitter circuitry rises to its resonance frequency, then dies away as the key is released. On the air this results in clicks or thumps at every keying, or in a slight change of pitch, resulting in a 'key chirp'.

All this is annoying to the ear, when a ham operator is straining to receive a distant signal. Worse, it can widen the band on the dial, so that the transmitter is occupying more ether-space than necessary. In the US, specifically, there are regulations dealing with these spurious radiations, since the FCC requires that

'The frequency of the emitted waves shall be as constant as the state of the art permits'. Hams use much ingenuity in devising filter circuits for transmitters to minimise these radio-frequency (RF) clicks, and the result can actually be seen if the receiver output (or transmitter monitor) is coupled to an oscilloscope, to display the waveform. If the wave-shaping process is carried too far, a 'backwave' can be set up, and the Morse Code signal may also be too 'soft', and difficult to copy (read) through the headphones.

Instead of simply modulating the CW with a continuous tone, then chopping it for Morse, we can have a microphone system handling speech or music and amplifying this electrically, then applying it to the CW train. In effect our comb-like pattern of CW is given the general shape of the audio-frequency envelope, which is what we do for 'A3' ham working, for 'fone' or voice communication. The main CW stream, though modulated, still acts as the carrier, and is at radio frequency.

When the carrier is modulated, sideband signals are produced on either side of the fundamental carrier frequency. These are known as the upper and lower sidebands, and are equal to the carrier frequency plus or minus the modulation frequency. Great economy of space and other technical advantages can be obtained if in effect only half the modulated CW stream is transmitted, thus eliminating either the upper or lower sidebands. This is known as single-sideband working (SSB). In practice the carrier is eliminated. With a normally modulated AM signal, about two-thirds of the RF power is in the carrier, and one-third in the sidebands; the transmitter power is therefore used to much greater advantage if part of the carrier and one sideband are eliminated. At the receiving end it is necessary to re-insert the carrier. SSB is used widely in commercial communications, but it is a field in which hams play a major role in circuit development. For example, Nick van Weede, G3VNC, has developed a solid-state transistorised SSB exciter using an RCA CA3020 linear integrated circuit, the total component cost in the UK being under £5.

With AM the modulation varies the amplitude (height) of the carrier CW. It would be possible to convey the same information by varying the frequency of the carrier a few kilocycles around its fundamental – in other words hurrying the carrier on, or holding it back, according to the imposed modulation. This is frequency modulation (FM). Frequency deviation of the carrier is known as the deviation frequency, and there are internationally agreed maximum values for this because, as will be realised, an FM transmitter would otherwise be 'wandering too far from its proper wavelength'.

In ham classifications 'F3' is voice communication on FM, or on its allied technique of phase modulation. 'F4' describes picture transmission of FAX (facsimile) by means of FM, and 'F5' is picture transmission by conventional amateur TV or by slow-scan TV. Just what these terms mean we shall see in the next chapter.

FM would be very useful for all ham communications if it were simply a matter of point-to-point communication but, as we shall see in Chapter Seven, long-distance ham communication makes use of reflections in the ionised upper layers of the earth's upper atmosphere, known as the 'ionosphere'. These reflections result in a slight frequency shift, which is obviously harmful to F3 and F4/5 systems, where the information is imparted by frequency change.

CHAPTER THREE

CQ-TV: AMATEUR TELEVISION

Today it has been forgotten that some of the world's first television transmissions were made by experimenters working as amateurs, or were transmitted on ham bands. Initial experiments by men such as Jenkins and Baird were conducted in the laboratory on closed-circuit systems, and later with 'televisors' and receivers linked by wire. But at the earliest possible stage they switched their vision experiments to actual radio transmission.

Only Dr Paul Nipkow (1884) was before his time, so could not make use of a radio link. At much the same period as Popov in Russia, he found that a scene or object could be optically scanned and converted into electronic pulses by a photo-electric cell – later to be re-created at the far end of the studio with a modulated neon lamp. Had radio transmission been possible then, the relatively coarse and flickering picture information could certainly have coped with the long-wave (1,500m and upwards) channels soon to become the basis of radio telegraphy; but as video experiments continued, and more picture information was produced, the transmissions had to take place on higher frequencies in order to handle their wider range.

C. Francis Jenkins of Dayton, Ohio, who was born of Quaker parents and spent his boyhood on a farm near Richmond, Indiana, came to Washington and eventually set up a research laboratory. In *Motion Picture News* (27 September 1913) he outlined a scheme for 'Motion Pictures by Wireless', and subsequently became a founder of the Society of Motion Picture Engineers, which ultimately expanded to become the Society of Motion Picture & Television Engineers – the major body in the USA for assessing technical progress and recommending TV standards.

In common with Korn, Amstutz and other experimenters, he was unable to use commercial radio channels, and for a time was driven to wire transmission. His first 'radio photographs' were witnessed by a *Washington Evening Star* reporter on 19 May 1922, and in October of that year Jenkins transmitted photographs over a desk telephone cable from his Washington office to the Navy Radio station NOF at Anacostia, DC. The US Navy was then using much the same frequencies as were eavesdropped on by many amateurs. Warren G. Harding wrote to him from the White House saying, 'The production of a picture in this fashion is certainly one of the marvels of our time'; and Herbert Hoover (then Secretary, Department of Commerce) told Jenkins, 'It represents a very startling development in radio'.

While Jenkins and lesser-known experimenters used amateur channels, only the great companies such as General Electric and Radio Corporation of America (today the RCA) could employ normal broadcast channels.

Elihu Thomson of General Electric Company put it on record that when he first saw Jenkins' prismatic ring (a circular prism used to scan the TV picture), he 'recognised it as the solution of a problem. It is perfectly possible, as you say, to employ this method of radio transmission of pictures on a very considerable scale, which would hardly be possible in transmitting them by the ordinary telegraph'. But General Electric did not offer Jenkins broadcast facilities in November 1922, and RCA was already embarked on its own TV laboratory programme.

In the amateur world the battle at that time was to be 'First Across the Atlantic', and 250 British experimenters took part in a not-too-successful test early in Febrnary 1921. The American amateur Paul F. Godley (2XE) sailed to Britain on the *Aquitania*, and with his 270m receiver set up in a tent at the very edge of the

Atlantic in Ardrossan, Scotland, picked up amateur station 1AAW clearly. During the following December transatlantic radio history was made in the reverse direction when 5WS, the Radio Society of Great Britain station, became the first amateur transmitter to be heard in the USA. The breakthrough in transatlantic picture transmission came on 30 November 1924, but it was an RCA achievement, owing little to amateur pioneering or frequencies.

RCA described the result as a 'Photoradiogram', when the national Press reported the transmission from London to New York. A transparent picture film was placed on a glass cylinder containing a light source focused in a narrow beam, and a GEC photo-cell transferred variations in light value into electrical variations. At the New York receiving end a paper drum printed out the result with 'a special vibrating fountain pen, drawn down by magnetic coils to record the picture much in the style of an artistic stippled engraving'. A significant feature of this RCA system was the use for the first time of the technique, now indispensable in any amateur or professional TV system, of sync (synchronism). In this first transatlantic picture transmission the constant pitch of a tuning fork was employed to maintain the necessary synchronism of the two drum motors 3,456 miles apart.

Amateurs pleaded for frequency channels well away from transmissions of this sort, and from the harmonics of arc transmitters that caused 'mush' to jam ham working; and more than 20 years later, when the BBC began the world's first high-definition TV service from the Alexandra Palace (London) transmitter, using Baird and EMI systems on alternate weeks, a more serious amateur problem manifested itself. By 6 February 1937, when the regular TV service was opened by the BBC, using the Marconi-EMI system exclusively, it was realised that this new public service was very liable to be jammed by amateur transmitters in the London region. The frequencies chosen for the pioneer 405-line system were 45MHz Vision and 41·5MHz Sound. The whole BBC-TV band then was within the third-harmonic range of amateur transmissions in the 14MHz band.

Amateur television has progressed to a state today in which DX vision can be conducted across the Atlantic through the OSCAR satellites, which were launched by US Service agencies, but were designed and built, and are now used operationally by ham enthusiasts on both sides of the Atlantic. An OSCAR can carry amateur communications, including a special form of amateur television of still pictures known as slow-scan TV (SSTV). The Services hold amateur radio in high regard, launching the sixth satellite in the OSCAR series in October 1972. In particular this one opened up new fields for ATV-SSTV.

To appreciate what the TV amateur is achieving, the non-technical reader may benefit from a brief pen-picture of television technicalities.

A scene, slide or film (telecine) cannot be transmitted in its entirety, but must be split up into scanned lines to provide signals for transmission. The more scanning lines per picture, the better is the definition; and the more pictures repeated per second, the less is the tendency to flicker. If pictures are transmitted and received at a 'frame-rate' faster than about ten a second, persistence of vision comes into play; each image stays on the human retina for a brief instant, and the brain records a moving sequence, not a flickering succession of pictures. For this reason early cinema projectors were run at 16 pictures per second, a rate raised to 24–5 with the advent of 'talking pictures', which carried the audio track on the film itself. In the television world we make great play about the number of scanning lines. Until the 1980s the British broadcasting service will be transmitting 405-line pictures on VHF, and 625-line pictures (colour as well as monochrome), and a slightly different 625-line system has been adopted in other areas. The United States has settled on 525 lines.

What is really more important, however, is the frame-rate, and the reason amateur as well as professional TV engineers largely disregard this is that there is not much they can do about it. For commercial convenience it is dominated by the local AC mains frequency, which in the United States is 60 cycles per second, and in Britain and other areas 50 cycles per second. Thus, whether the picture is scanned horizontally in the raster (frame) by 405, 525, 625, 819 lines (adopted in France through the 1960s) or 1,000 lines (employed in specialist military and scientific systems), the frame-rate is almost invariably tied to the mains frequency. It provides a nationwide means of synchronisation in the vertical direction.

When Baird, Jenkins and the other pioneers were scanning their TV objects with spiral rows of lenses mounted in spinning discs, it was generally convenient to lay out a spiral of thirty lenses, giving 30 picture lines at each revolution. As the discs revolved at $12\frac{1}{2}$ revolutions per second, there was noticeable flicker. Some amateurs begin their home TV experiments with mechanical systems like these.

The first big forward step came when Vladimir Zworykin brought the first workable picture (kinescope) tube to the world market, and pictures could be scanned electronically instead of mechanically. In theory there was now no limit. John Logie Baird increased his scanning rate to 240 lines per picture, with a repetition rate of 25 per second. Then the Marconi-EMI physicists introduced 'interlaced' scanning, still further reducing any possibility of flicker.

Naturally the eye could not follow the electron spot as it painted the Baird 240-line picture, but it tried to do so. This became obvious when the Baird system was broadcast experimentally from the Alexandra Palace transmitter. With Marconi-EMI the final result is a 405-line picture scanned 50 times a second; but instead of tracing out the whole 405 lines at one go, the spot is made to cover $202\frac{1}{2}$ lines in $\frac{1}{25}$ second, and it then returns to the top and in the next frame covers the alternate lines. This system of interlacing has become universal for professional broadcasting. Many amateurs adopt it, and it is also used for the higher-quality commercial closed-circuit TV.

Of course it is not sufficient to draw out this 405-, 525- or

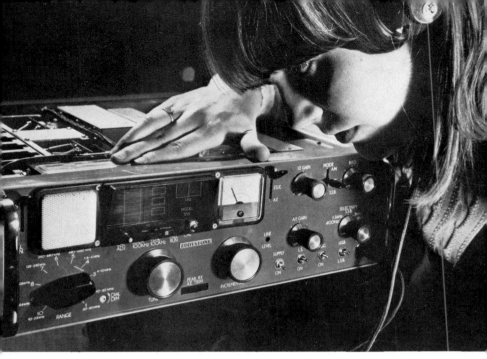

Plate 9 Typical wide-range general-purpose communications receiver of the 1970 era – the Eddystone Radio EC 958 – made by a British manufacturer in the GEC-Marconi Electronics group. This rig has an optical projection scale, and ranges from 30MHz to 10kHz in a continuous sweep

Plate 10 Shortwave link to the rescue. A US ham works on the emergency service during the South Dakota floods of 1972

Plate 11 Dutch amateur Dick van Breen, PA0FX, operating the International Amateur Radio Union station PA6IARU at Scheveningen, Holland

Plate 12 Keen scout members on duty at a Jamboree-on-the-Air. The DX Century certificate is displayed above the receiver

625-line raster. The spot is drawn across, up and down by electrostatic plates or coils surrounding or actually in the electron tube, and two separate synchronising (sync) signals are given – one the 'line sync' pulse and the other for 'field sync'. In addition to moving, the beam has to be modulated to convey light and shade and all the myriad intricate pictorial details. This is one reason why TV amateurs find ATV such a fascinating hobby. It embraces not only electronics but artistic appreciation, together with picture composition and lighting.

Amateurs as well as professionals need to examine the transmitted waveform, which is very complex, with extremely minute time factors. On the basic 405-line system, for example, the electron beam takes just 98 microseconds to trace out the lines of a single frame – and in this tiny fraction of time it is modulated to paint light and shade. About 18 microseconds are taken up while the spot returns to the start again, and this is known as 'flyback' time. During this blanking period there is an extremely short portion of the waveform called the 'front porch', which assists receiver circuits tuned to the transmission, and this takes place in about 2 microseconds. There is another little interval in the waveform after the line pulse, and this is known as the 'back porch', which gives time for flyback. Synchronisation pulses are produced by an SPG (sync pulse generator), and these pulses are actually transmitted in the tiny intervals between 'picture painting'. They are really extremely accurate and stable electric clocks, and their design and operation take up a good deal of amateur-TV time and patience.

At the transmitting end of the TV chain the scene is focused by a conventional lens system (professionally with turret lenses or zoom) and brought to the face-plate of the camber tube. At the other end of this tube the radiated electron beam is being directed back and forth by the SPG and the mains frequency, covering the small active rectangle of the face-plate with 405, 525, 625 or amateur non-standard lines per second, and at repetition rates of 50 or 60 per second.

There are many types of camera tube, but basically only two

forms: the vidicon, in which the light falling on its face-plate target results in variations in conductivity (hence 'photo-conductive'), and the 'photo-emissive' (image-orthicons are of this type) in which the beam of light falling on the target generates a voltage dependent upon the brilliance. The photo-conductive vidicon is the basic cheapest tube, largely used by amateurs. The NV Philips Plumbicon is a very sophisticated lead-oxide-target vidicon, and the English Electric Leddicon betrays by its very name a somewhat similar target type.

Voltage and/or current variations from the tube are amplified and transmitted, with the complex sync pulse, and of course with an allied audio channel where a microphone accompanies the televised scene. At the receiving end the electron beam in the receiver's cathode-ray tube (CRT) is caused by the sync pulse and mains frequency to follow the movement of that in the camera tube exactly. Where frequency bands must of necessity be restricted, as in the amateur world, the fact that such a huge parcel of information has to be handled is a nuisance.

In practice the satisfactory band of the audio channel can be limited to some 10,000 cycles per second, but on, say, a 625-line TV system the band needs to be 500 times as wide, or some $5\frac{1}{2}$ million cycles a second. Because of the lower visual definition possible on 405 lines this occupies less 'air-space'. It needs about $3\frac{1}{2}$ million cycles per second instead of $5\frac{1}{2}$ million, so we say that this takes up a 3MHz band.

The aim of amateurs using the slow-scan techniques is to send pictures on a channel occupying only 2·8kHz width. Although this conveys all the necessary picture information from black through shades of grey to white, it demands no more air-space than a normal single-sideband (SSB) transmission. In practice it is not necessary to use SSB. A conventional FM or even AM rig can be modulated for SSTV.

Of course the end result of SSTV is a low-resolution still picture that is formed in approximately 7·2 seconds, so for recording it amateurs generally use a camera coupled to the monitor. There is a relatively simple layout in which a Polaroid-

Land camera is set up facing the monitor, with its shutter open for the full 7·2 seconds. The developed print is of course available a few seconds later. History was made when, in the early 1970s, slow-scan transmissions by Copthorne Macdonald (WA2BCW) were received in Britain by G3LEE and G3AST, and many Polaroid-Land recordings of transatlantic SSTV have since been made.

The whole development of television throughout the world has been a fascinating quest towards technical perfection, entailing the development of VHF and UHF transmitters, links and relays, scanning and video-tape and video-disc recording systems, and a professional ergonomic approach to what is one of the most sophisticated electronic services in the modern world. So it is little wonder that amateurs as well as professionals in their off-duty hours like to undertake their own experiments. The amateur TV movement began to grow in the late 1940s, inspired no doubt by ex-Service radar workers and the availability of surplus radar equipment that could be converted to ham-TV use.

Regular publications, including *Wireless World*, the *Short Wave Magazine*, *QST*, *CQ* and *Ham Radio Magazine*, began dealing with ATV progress, and the ARRL and RSGB took this in their stride. After all, it was yet another type of signal to be handled on regular amateur bands.

In the USA Ron Cohen (K3ZKO), Philadelphia, Penna, had a thriving ATV club going by 1971. For several years he directed the club, and produced *A-5 Amateur Television Magazine*, doing most of the photo-typesetting on his own secondhand IBM Executive. In all this he was aided by Al Lipkin (W3AEH) as associate editor.

Lack of capital was naturally one of the teething troubles, but TV enthusiasts were soon complaining that other branches of the electronic industry were taking their new recruits. Ron Cohen himself complained in 1973:

Looks like jobs in the RF field are plentiful at this time. Seems like companies cannot find people to fill jobs in the two-way [communication] shops. I guess the schools are pushing the digital end of electronics more these days. I even had to turn down a job in the Virgin Islands. The job was on ship repair of radio and radar, and I get sick even being tied up in port.

TV amateurs began to find it profitable raiding ship-breakers' yards for klystrons and other ex-USN equipment useful for the RF end of ATV. As 'Captain Video' of the US Amateur TV Club, Cohen led the campaign encouraging the ATVs to 'get RF' – that is, to transmit on radio, not confine themselves to CCTV (closed-circuit television) – and to build their own rigs instead of merely modifying secondhand professional equipment. Commercial surplus gear hardly constituted a hobby.

'Stop running closed-circuit TV around the house', Cohen urged. 'So many hams believe that the RF comes last and the camera first. Get the RF on, then you will really be in touch with the ATV-ers in your area.' As for frequencies, he began warning back in 1972: 'The band known as the 450 [70cm, 432 or the $\frac{3}{4}$m band] is in deep trouble. We the amateurs will have to self-regulate the band so that each of the different modes – AM, CW, SSB, FM, ATV and satellite – will not interfere with each other. The FCC will step in and do it for us if we don't straighten ourselves out.' It must be remembered that the Federal Communications Commission's rules are not meant to be broken, and a violation can bring a fine of up to $500 for each day it lasts.

In a rather clumsy attempt to regularise amateur radio, the FCC 'leaked' a threat in 1973 that all ATV equipment and much non-standard ham gear would have to be type-accepted. Since those days it is true that SBE Linear Systems, Cush Craft Corporation, Venus Scientific Inc, and other specialist companies in the USA have produced commercial accessories for ATV; but when the FCC's possible intention was announced, an ATV transceiver such as the Motorola 744 cost around $40 and a camera $100. Cohen asked:

What is ham radio becoming? It does look like the FCC is pushing the hams along with statements implying that all ham equipment should be type-accepted. Just think what would happen with ATV. Not many stations would be operating, as there is no commercial equipment on the market. You would still be able to homebrew and modify surplus equipment, but this would have to be packed off to the FCC for evaluation and a stamp of approval. If lucky, you might have to wait only six months for the return of the equipment.

A complete list of amateur-band channels open to ATV in the USA is published by the ARRL (see *The Radio Amateur's License Manual*), and certain policy difficulties, such as the fact that it is not a straightforward matter to monitor ATV transmissions, have been overcome. As a general principle, it is held that the FCC monitors should be able to 'understand' what amateurs are transmitting and, if necessary, view the transmitted picture. For much the same reason British TV amateurs are not permitted to transmit 525-line pictures, as the British Post Office monitors are said to be capable only of handling 405- and 625-line pictures. On the reverse side of the coin, the FCC has tightened up regulations covering TVI and BCI (TV and broadcast interference). Forfeitures or small fines up to $100 (or $500 in the aggregate) can now be levied on amateurs on an administrative basis without the need for FCC to go through the Federal Court system. Such fines are levied, of course, only for repeated and/or wilful violations. Penalties probably equally as hurtful as fines are the 'quiet hours' the FCC can impose, usually between 2000 hours (8.00pm) and 2230 local time, or between 1030 hours (10.30am) and 1300 hours on Sundays. Amateur transmissions of radio, phone and ATV may be cancelled during these 'quiet hours' until neighbourhood complaints cease.

With the growth of SSTV and other VHF-UHF-TV activities in the USA, the amateur television movement became busier and more complex. Currently the *A-5* journal has been handed to a new editor and publisher – Henry B. Rhu (WB8HEE), Whitmore Lake, Michigan.

The British amateur television movement is senior to that in the USA, and indeed there was historic cooperation between British and American hams long before the launching of *A-5*. On 22 November 1959 slow-scan pictures from WA2BCW in New York were received by G3ASR in Yeovil, Somerset, England. The British Amateur Television Club (BATC) was founded in 1949. It is not a section of the Radio Society of Great Britain, but of course is affiliated to it, and currently the membership tops the 1,000 mark.

In the United Kingdom, unlike the USA generally, amateurs are recommended to use a video waveform similar to that used on-air by the BBC and the Independent Broadcasting Authority. This means that a domestic TV receiver can be used to handle amateur transmissions by the simple addition of a slightly modified UHF tuner unit.

The following channels are currently available for professional TV work in the UK:

Band I (VHF) 41–68MHz, 405-line transmissions of BBC 1
Band II (VHF) 87·5–104MHz internationally, but limited in the UK to 87·5–100MHz, and used chiefly for VHF radio, including stereo, national and local BBC stations, and IBA local radio
Band III (VHF) 174–216MHz, channels 6–13, 405-line TV transmissions by BBC and IBA
Band IV (UHF) 470–582MHz, channels 21–34, 625-line colour and monochrome BBC1, BBC 2, IBA and transmissions planned for the future
Band V (UHF) 614–960MHz internationally, but limited in the UK to 614–854MHz; used for channels 39–68, 625-line TV on BBC 1, BBC 2, IBA and future transmissions. (As with Band IV, a fourth programme has been catered for in these frequency allotments)
Band VI (SHF, super high frequency), satellite-TV and radio, and for terrestrial links

BATC members in the UK need to obtain an amateur TV transmitting licence (currently £3 a year), and the applicant must have a pass in the Radio Amateur examination (theory), but no knowledge of the Morse Code is required. Operation is permitted in the 70cm band and on shorter wavelengths. SSTV is the branch of the hobby gaining worldwide popularity, and in the British Isles it may be transmitted also in the 40m, 20m and 15m bands by holders of Class A licences. The supplementary permit is issued free on application to the Ministry of Posts and Telecommunications, London. British amateurs have found it possible to get very satisfactory SSTV pictures with the aid of tubes having long persistence (the glow pattern staying on the tube face for an appreciable period of time) and with a raster of 120 lines. With 50Hz mains supplies in the UK, this means a line frequency of 16 2/3Hz, and one frame in 7·2 seconds. In the US and other 60-cycle areas the SSTV standard is 15Hz, and one frame in 8 seconds. For amateurs the one big advantage of this still-picture system, in addition to its being handled over conventional HF channels, is that, with a 3kHz bandwidth, it can be recorded on a conventional domestic cassette or reel audio tape machine.

Satisfactory results can be obtained with inexpensive equipment in every field of ATV. The BATC shows its members methods of constructing a flying-spot scanner using a 5FP7 CRT and a 931A photocell for about £10. This simple basic system will handle positive or negative transparencies.

The *British Amateur Television Club Journal* is edited by Andrew Hughes. The current Club president is R. S. Roberts (G6NR), and the publications manager is Malcolm J. Sparrow (6KQJ/T, G8ACB). The general secretary, Joe Rose (G6STO/T) is one of the pioneers, and other members who have played a major part in the British amateur-TV movement since the early days include John T. Lawrence (GW6JGA/T, Flintshire), Ian Lever (G8CPJ, Kent), and C. Grant Dixon (G6AEC/T, Herefordshire).

The BATC performs a useful service in making club equipment available to members at cost prices, current examples being English Electric P849 amateur-grade camera tubes at £11·55, 2–3in diameter amateur-grade vidicon tubes at £11, and ex-studio vidicons at £6·05 when available. Even the old-type broadcast 4½in image-orthicon tube (9565 and 9564) is sometimes available to members around £11.

Examples of the normal run of equipment exchanged between BATC members include a Pye 2780 14in monitor at £10, a Marconi line clamp amplifier at £3, a Marconi sawtooth waveform generator at 75p, a Pye 'window' generator at £2, and a Rank Bush Murphy CV101 colour monitor with NTSC decoder at £15. It will be seen that amateur TV necessitates a good deal of grey matter but not much money.

It seems only yesterday, but in fact it was in 1949, when Michael Barlow wrote letters to *Wireless World*, the *Short Wave Magazine* and other journals, and succeeded in contacting a few British amateurs interested in ATV. The first issue of *CQ-TV* came out in October 1949, in duplicated form and comprising only twenty-five copies.

Initially almost all amateur UK TV signals were generated by flying-spot scanners, as camera tubes were expensive and difficult to obtain. The 5527 iconoscope required a Board of Trade licence, and cost £27·50 (more than the average week's salary then, for a professional man); furthermore, it was not very sensitive, and gave a picture that would be considered definitely below par by modern amateur standards. A few lucky amateurs, however, acquired a tube, notably Ivan Howard (G2DUS), who gave the first public demonstration of amateur TV under the banner of the BATC in April 1950. A little later, as a result of on-air tests (in which the BATC cooperated) to see if ATV would interfere with certain commercial and military transmissions, the GPO (later the British Post Office) decided to grant TV licences to amateurs, and the TV ham had become a reality. Under the first chairmanship of C. Grant Dixon the club expanded, and within 5 years BATC members were being enrolled from Eire, France, Finland, Holland, Germany, the USA, Canada, South Africa, Australia and New Zealand. Two-way working was possible between only a very few of these countries (eg USA/Britain in 1959), but all wanted to share in the pool of growing BATC experience.

In May 1952 the first two-way amateur TV contact in the world was achieved, between G3BLV/T and G5ZT/T. In August 1953 G3GDR received G2WJ/T at 34 miles range with only 2W peak input, and in the December of that year amateur colour pictures were produced in the UK for the first time. Colour pictures were transmitted, two ways, over a 12-mile path (Great Baddow/Dunmow) in the Spring of 1956, and around the same week another record was set by BATC members in exchanging monochrome signals over a 38-mile path. Following the epic transatlantic exchange in November 1959 (WA3BCW New York to G3AST in Somerset), the two hams G3ILD and G3NOX/T showed in the autumn of 1963 that two-way pictures could be exchanged over a path in excess of 200 miles.

The BATC holds a convention once every 2 years, at which members display their equipment and exchange ideas, and when the first meeting was held in 1951, the fact that twenty-five members could attend was considered remarkable. At a subsequent convention the British company EMI Electronics (partners with Marconi's in transmission aspects of the world's first broadcast-TV system) made available six Emitron camera tubes to the club on a permanent-loan basis. This generous move resulted in a very considerable boost to the amateur-TV movement: one of these tubes was displayed in a frame-sequential colour camera at the 1955 Convention, and history was made the following Spring when colour test patterns were transmitted over the air by amateurs. This was before the BBC had started its colour tests. Again, in the autumn of 1966, the BATC convention laid on the first demonstration of a three-tube colour camera built by Michael H. Cox, an amateur who learned his TV technicalities with (British) ABC-TV, and whose 'Coxbox' is today an integral part of many international colour-TV broadcasting stations. It produces colour images from monochrome originals.

Britain decided to adopt the PAL (phased-alternation-line) colour-TV system instead of the American RCA-inspired NTSC (National Television Standards Committee), but it was a British amateur, G6ACW/T, who in April 1968 contacted G6LEE/T and proved that amateur PAL colour transmission was possible on 70cm.

A-5 and *CQ-TV* have triggered off amateur television interest in almost every country. The Japanese *JARL News* (printed entirely in Japanese characters) follows the American amateur-TV scene closely, but in continental Europe Belgium and Germany have led with organised club interest.

Willy Everaert (ON4WM) is editor-in-chief of the journal *ATA International*, the organ of the Amateur Television Association, an international non-profit group established at De Pinte, Belgium, in 1967. It is devoted to fast- and slow-scan TV, facsimile (such as weather-chart) and amateur video communication in general. As an international journal, it is printed almost entirely in English, and amateurs in the US, the UK and other countries find it well worth while enjoying membership, since the ATA helps them find rare components. *A-5* is of course US-orientated ('Pixe-Verter Kit adds RF output to video-only-type cameras and VTRs, $6.95'), and *CQ-TV* tends to be overloaded with Marconi, EMI and Pye equipment in its advertising section. On the other hand, the ATA may be able to unearth a 70cm ATV UHF converter from Dortmund, or a Philips Plumbicon tube 55876 from West Germany.

The current president of ATA is Willy Van Marck (ON4RT and the secretary Erik Platteeuw (ON4LP). In conjunction with the BATC in Britain and the AGAF (Arbeitsgemeinschaft Amateurfunkfernsehen) in West Germany, the ATA has run a number of successful international amateur-TV contests, covering such tricky matters as two-way dual-band visual contact (eg transmitting vision on 435MHz, and receiving on 1,296MHz) and so scoring 4 points per kilometre. Even portable TV transmitting stations have taken part in these high-prestige contests.

The AGAF's own journal, *Der TV-Amateur*, is probably limited in interest outside Germany, since it is published entirely in German, but test cards, block diagrams and the latest list of frequencies, transistors, stations and other data do not call for translation.

CQ-TV: AMATEUR TELEVISION

The amateur-TV movement has been worldwide since February 1973, when CQ *Elletronica Magazine*, Bologna, organised the first really effective World SSTV contest, 3·5 to 28MHz. The ATV movement has been growing in Canada since September 1973, when George Davis (VE3BBW) and Tom Atkins (VE3CDM) founded the Ontario ATV Association for fast- and slow-scan TV. In New Zealand Douglas Ingham (ZL2TAR), Lower Hutt, has been transmitting PAL-coded colour since 1972.

In SSTV the BATC broke new ground with a 1972 first edition of a 16-page explanatory paper by B. J. Arnold, MA (G3RHI), followed by a modified and extended edition in November 1974. The American *Slow Scan Television Handbook* by Don Miller (W9NTP) and Ralph Taggart (WB8DQT), which is available through the ARRL, RSGB, *A-5* and *CQ-TV* (BATC), covers ATV hams on both sides of the Atlantic.

CHAPTER FOUR

JAMBOREE-ON-THE-AIR

Radio amateurs organised in the ARRL or RSGB who regard themselves as pioneers in international communication would be surprised to learn they were forestalled, over half a century ago, by the 1st Arundel (Sussex, England) Troop of Boy Scouts. In 1912 a handful of these Arundel Scouts went to the department store of A. W. Gamage Ltd in Holborn, London, and in conformity with Post Office regulations entered the details of their station when they were buying components for their amateur rig. In those days there was, of course, no ready-made gear. Amateurs even wound their own coils and assembled their variable 'condensers' (capacitors).

This troop, which was stationed at the Swallow Brewery, Arundel, registered its call-sign as XBS, for this was before the 'Ws', 'Gs' and other prefixes came in. The troop's 50W rig was powered by accumulators (storage batteries), operated on 200m, and was stated to have a sending range of 5 miles and a receiving range of 800 miles.

The Gamage directory entry, now a historic amateur document of museum status, has a scribbled addendum: 'Also licensed for portable apparatus to work within 5-mile radius most Wednesday and Saturday afternoons, and other times not fixed.' In other words, it was a mobile, operating in the pioneering years of the international scout movement.

Today the Jamboree-on-the-Air (JOTA) has become an official event of the annual scout calendar, to such an extent that

JAMBOREE-ON-THE-AIR

many scouts in various countries mistakenly believe that radio scouting is restricted to JOTA. A glance at the early history of radio scouting, however, puts the whole project in a true light. All that is generally known among scouts is that this section of the world movement began in Britain in 1957 when a number of scout radio hams clubbed together in Sutton Coldfield to hold a 'ham-fest'. The suggestion was made, and enthusiastically adopted, that scouts all over the world should try to contact each other on a fixed day every year by means of amateur radio. It must be remembered that signalling is part of scouting skills, and there are proficiency badges for knowledge of radio. It is all part of the basic Baden-Powell (1857–1941) philosophy that each scout should be 'a brother to every other Scout'.

Comparatively few scouts are able to get to the great World Jamborees held every year, partly because of the expense and partly because of political or other national barriers. Naturally, those who can attend the Jamborees are fortunate indeed, because nothing can compare with the experience of camping in a foreign country and meeting and making new friends among the thousands of scouts there from all parts of the world. The Sutton Coldfield meeting realised this, and knew that by means of amateur radio it might be possible for scouts to meet and talk – on the fixed Jamboree day – without leaving their home towns. An added benefit from this annual scouting event is the fact that several scouts have gained amateur licences as a direct result of being introduced to the hobby at a Jamboree.

Nowadays the event is controlled by the World Scout Bureau (Bureau Mondiale du Scoutisme) in Geneva, L. F. Jarrett (HR9AMS) being the Director of Administration and JOTA Organiser. It has become virtually a scouts' radio League of Nations, and JOTA information has to be published not only in the 'official' languages of English and French, but in Portuguese, Norwegian, Icelandic, Finnish and many others.

JOTA owes its inception to Leslie Mitchell (G3BHK), who joined the Scout movement in 1936 but whose one chance to attend a World Jamboree was spoiled by the outbreak of World

War II. Instead, he volunteered for the Royal Navy, which taught him a great deal about radio theory and practice, as well as the Morse Code, which of course is the golden gateway to full experience as a radio amateur. As a result, he became G3BHK, one of the first immediate postwar British 'G's', in 1946, and 3 years later he was running his first-ever scout radio camp. As signalling has always been a vital part of scouting it is rather strange that the movement did not get around to it earlier. Mitchell has said:

> We had about twenty scouts really interested, but in that camp it took us all the weekend to work some sixty stations. We operated mostly on 80 metres, and the transmitter was crystal-controlled. The most popular crystal frequency in those days was 3740kHz.
> Following this camp, we ran several scout radio events through the years, but very little interest was shown internationally. At that time you could have counted on the fingers of one hand the scout amateurs in the United Kingdom – in fact we could not raise enough scout hams to man the station GB3SP of the 1957 World Jamboree at Sutton Coldfield. Therefore we were forced to ask the local amateur radio club to run the station and, whilst it was a magnificent effort, the truth was that only two operators in the station were in scout uniform.

It was discovered, nevertheless, that there were some ten scout amateurs attending the Jamboree from other countries. Groups used to meet over coffee each morning and discuss their international experiences as scout amateurs and as short-wave listeners. It was during these coffee sessions that someone suggested they might put aside one day a year to contact all scouts on the ham bands, and Mitchell was asked to organise such an event. He explains what happened next:

> On returning home from the Jamboree I realised that although this venture sounded exciting, it was doomed to almost certain failure. There were so few of us that contacts would be very sparse, if at all: and I feared that interest in it all would quickly wane. That's when I hit on the idea of inviting amateurs throughout the world to help the scouts, encouraging the scouts themselves

to help run ham stations and to witness all the aspects of the operation.

In November, 1957, we ran a pilot station, a 40-watt AM transmitter from Fielden. Local scouts attended. In twenty-four hours we made contacts with 63 stations in some thirty countries, and the lads found it so interesting they were impatient for another try. From our experiences over that weekend the rules were drawn up for the first-ever Jamboree-on-the-Air in 1958. But our results then were modest.

The *Short Wave Magazine* of July 1958 reported:

SCOUT JAMBOREE-ON-THE-AIR. Some 40 overseas stations were known to have been on for the first international Jamboree-on-the-Air during the weekend of May 10th/11th. As it turned out, conditions on the DX-band were poor . . . Not many of the overseas stations were worked from the UK. About twenty UK scout stations joined in and worked one another, mainly on 80 metres. Scouts operating on the G prefixes were successful in working other scout groups in DJ, F, OE, OH, ON, SM, ZE, ZS, W and ZL, and even a JA station with a YL Girl Guide in the station. [In clear, this means contacts with Germany, France, Austria, the Åaland Islands, Belgium, Sweden, Rhodesia, Republic of South Africa, the USA, New Zealand and Japan.] Most consistent and successful of the UK participants were G3SP in Sutton Park and G3BHK in Reading. It is hoped to make this an annual event, as in spite of the disappointing DX conditions a great many interesting scout contacts were made.

But the Jamboree was conducted over the entire globe, and in the *World Scouting Magazine* there was this report:

Keen interest was shown in the Jamboree-on-the-Air, but not, unfortunately, by the Clerk of the Weather. Reception on the DX bands this weekend was the worst for many weeks. In Ottawa, where the International Bureau was invited to operate from Ray Thornton's station VE3RT [Province of Ontario] conditions were so bad that we could pick out only five stations – four in Canada and one in the United States. We believe, however, that our calls were heard in many other parts of the world. The following letter from Pat in Victoria, British Columbia, proved that the Canadian West Coast heard us: 'I am sending this note from

Bell, who is blind. He is a short-wave fan and belongs to the Boys' Life Radio Club of the Boy Scouts of America, in New Jersey. He was listening-in on your Jamboree over the weekend, and picked out VE3RT Ottawa, and G3BP, Gilwell Park, in England.'

The World Bureau assisted in publicising the event, and, despite the poor conditions at the outset, it proved, as Mitchell had always believed, that the weekend of world contacts would prove to be the best advertisement. This 1958 experience, nevertheless, showed many deficiencies in the organisation, not least that it was far too much for one man to do; and so at the end of May 1958 the World Bureau was asked to take over JOTA officially. This it did willingly, realising the tremendous possibilities for such an event, and the Jamboree-on-the-Air became an official event of the annual scout calendar.

JOTA has progressed far since 1958, but old stagers recall the pioneering occasions with nostalgia. As one said: 'It was very exciting in the first few years because it was frankly unusual to hear a scout station on. And when you finally did, there was keen excitement until you made contact. After that, the excitement was even higher. It is all different now. During the JOTA weekend one gets a jolt on hearing anyone who is *not* a scout station.'

A glance at the numbers of participating countries shows that the figure has stabilised around seventy, after a sharp increase in the first years. The number of scouts taking part has also steadily increased, from, for example, some 100,000 in 1970 to just under a quarter of a million in 1972. These are only approximate numbers, but it is certain that more scouts than ever do now have the opportunity to speak to and listen to brother scouts all over the world. JOTA is established as an annual event on the amateur radio bands, just as are the large contests of clubs and non-scouting amateurs.

Important points were brought out in 1973 in Oslo at the Nordic Conference on Radio Scouting, organised by the Norwegian Boy Scouts Association and the Camp Committee of NORDJAMB-75. The camp at Sollihøgda, some 30km west of Oslo, extended hospitality to fifteen participants and eight

40 years of amateur progress:

Plate 13 Bernard A. Barton, G2HFB, of Boston, Lincolnshire, who began as a ham by constructing a Cossor Melody Maker from an amateur blueprint in the 1930s

Plate 14 Picture taken off the tube from one of the first amateur-TV transmissions across the Atlantic, via the satellite OSCAR-6

Plate 15 Member of the Racal Amateur Radio Club operates G3RAC at an RSGB National Field Day HF contest, using the latest Racal RA 1772

Plate 16 Grant Dixon, G6AEC/T, of Ross-on-Wye, operates a typical advanced British amateur television station

Plate 17 Jim Johnson testing a 10lb transmitter at the RCA Space Center, Princeton, NJ, before it was sent more than 220 million miles to Mars to gather scientific information. The next generation of hams is looking forward to getting DX QSL cards for signals bounced off planets.

observers, including the following: *Denmark*, Arne Gotfredsen OZ3AG, Arne Herkild OZ3YE, and Kai Stecher OZ2YS; *Finland*, Vilhjalmur Kjartansson TF3DX; *Sweden*, Ernst Andersson SM4BMX and Per Carlson; and *Norway*, Cato B. Almnes LA9PF, Arild Delphin LA2DQ, Christian Dons LA5OQ, John Fr. Klepzig LA7IL, and Thorbjorn Pedersen. Observing amateurs included LA2ZR, LA2NL/G5AXR, LA8IP, LA2SR and LA1SP. Programmes of radio scouting in the five Nordic countries were fully explored, and, as a result, the outcome of the conference was of permanent interest to scouts in other world zones.

For example, it came as a surprise to find that, like scouting itself, radio signals know no boundaries. There are millions of scouts in the Iron Curtain countries, including Poland, Czechoslovakia and Hungary. While these are not officially recognised by the Boy Scouts World Bureau, everything is done by radio scouts to promote friendship with radio contacts in these countries. During one Jamboree-on-the-Air an ex-scout farmer just on the other side of the Iron Curtain offered (in Morse) to allow British scouts to camp in his field – if they would come across to Czechoslovakia!

A valuable warning was given by Les Mitchell at a World Jamboree in the early days of JOTA:

> We are all here because in the first place we are scouts. We must avoid the impression that we are trying to turn every other scout into a little radio genius. To be candid, a scout can be very deeply interested in Jamboree-on-the-Air without wishing to know what happens under the lid of the transceiver. In the same way, you do not need to know how to build a motorcar in order to be a good driver. Our objective is: 'How can we improve our *scouting* by using the facilities of amateur radio?'

Potential short-wave amateurs who have little interest in the world scout movement will feel some sympathy with five points that came out at early JOTAs, because they apply to enthusiasts, young and old, who know nothing of scout disciplines. The problems include (1) The rather specialised ham language, (2) microphone shyness, (3) speech quality of contacts, and (4) the contest attitude.

Codes, abbreviations and a guide to ham language are given in the Appendix to this book. Some scouts and similar groups of youngsters try to enlist the aid of an additional experienced amateur whose sole responsibility is to translate and explain, while other groups have explanatory hand-outs of abbreviations, the Q code and other 'slanguage'. Many find that a visit to a busy radio ham before taking part in the JOTA is advantageous, for they can then see for themselves that most of this ham language and the code abbreviations are essential, either to save strain on the wrist for Morse working or to save air-time generally.

It will seem odd to adult operators that scouts may suffer from microphone shyness, yet if you hand the public-address microphone to anyone without experience, the most they can usually think of as a test-piece conversation is 'Hello, Hello, one, two, three, four . . . Testing, testing!' This embarrassment soon fades away, however, and the main enemy of every good fone-contact on the air is the thoughtless microphone hog. The use of tape recorders helps scouts to be less conscious of their own voices, and, as Cato Almnes (LA9PF) pointed out at a JOTA: 'The conclusion is that if the boy knows what to talk about, the shyness vanishes.'

With so much inferior audio quality in everyday broadcasting, particularly with the all-too-frequent 'phone-ins' where the speech band is so limited on cable, people of all ages are becoming accustomed to deciphering for themselves what others are saying, as if down an echoing pipe. Nevertheless, speech quality is a problem. Many groups of scouts likely to take part in a JOTA are given introductory courses, so that newcomers can become used to the rather odd SSB and DX sounds. Just what can be done depends to some extent upon the country of origin. In the USA, for example, citizen-band radio provides some experience in frequency-limited speech. The trend is to take 2m, 70cm and the citizen band into use for pre-JOTA practice.

Point 4, the contest attitude, is one that arises not only in radio scouting, but wherever amateur radio is linked with a public movement. Naturally each JOTA tries to link together as many scouts as possible, for the benefit of scouting, not to set up new DX records.

The JOTA idea was born at a camp. For the following few years scouts were almost compelled to operate from a friendly ham's shack (radio room) during the event, but in recent times the picture has changed. Scouts have their own transmitters, and enthusiastic licensed amateurs take their gear to the scouts. This cooperation can be practical. Who is better than a scout to help erect an antenna tower up to 100ft or more? There are scouts participating in SSTV (slow-scan television), as well as a sport peculiarly linked to scouting, and known as fox-hunting. Licensed amateurs in a group of three or four hide in various locations. Competitors use receivers ('fox traps') with directional antennas to take bearings, and so try to find the foxes. Of course one does not need to be an experienced ham to operate a fox trap, and there are scouting publications in most European languages giving construction details of suitable short-wave traps. In a really busy jamboree each participant has a battery-operated fox trap, some detailed instructions, and at the 'Off' he starts out after the fox. At least two traps are needed to get a cross-reference, as in any more sophisticated form of direction-finding and spotting.

It is a game, but a sport leading to more serious applications of radio scouting. It emphasises aspects of amateur-radio activity as a permanent distress and emergency communication system, ready at all times for police, fire and other emergency work. In various European countries and in Latin America police authorities are in touch with local amateur groups; but in others (including Great Britain) there are legal restrictions on this sort of use of amateur working.

Records of many kinds have been set up and broken through the years at the JOTAs. For example, the thirteenth World Jamboree was held in Japan, where station 8J1WJ was almost a jamboree in itself, owing to the efforts of Japanese scouts and leading manufacturers. No fewer than nine separate stations operated simultaneously, on all bands and in all modes, including teletype and facsimile weather pictures. A real 'antenna farm' surrounded the encampment, with two full-size rotary beams for the HF bands, another for VHF, and innumerable dipoles. Japanese scouts willingly stood aside and let foreign visitors take over the ham station. On this occasion a severe typhoon hit the Jamboree, but foreign visiting scouts maintained a round-the-clock operation while others held down the flying canvas to stop the torrential rain from ruining equipment, and were also busy through the night digging trenches to drain the water away. All this is good scouting, and of course fine ham team spirit.

At many a JOTA difficulties have been experienced contacting stations outside the Asian Pacific zone, because of QRM (interference), and scout jamborees in fact are a practical way of conducting world-wide investigation into QRM, bands and optimum times. The JOTA programme has become so well accepted in the USA that not all scouts and hams trouble to contribute reports from the mainland, despite regular good results from WØ6BSA and W17BSA, while the national headquarters in New Jersey is always active under its special call-sign of K2BSA. In remote Okinawa short-wave communication is excellent. The Okinawa Amateur Radio Club supports radio scouts, and on one recent JOTA three Cub Packs, five Boy Scout Troops and eight Girl Guide units were able to participate in contacting 203 stations in fifty-seven different countries.

Radio amateurs confined to their own shacks envy the transportability of radio-scout equipment at JOTAs. On one occasion, when Alan Ritchie (ZS6XK) was national organiser for South Africa, he suddenly received a report that Mafeking station ZS6JAM had no equipment. Obviously Mafeking had to be relieved! Ritchie cancelled his own arrangements to transmit from the local scout hall, and on the Friday evening left Johannesburg for Mafeking, 175 miles distant, with a transceiver, linear amplifier, dismantled beam antenna, 30ft mast, cables, meters and a 1·5kW generator in a station wagon, followed by a scout leader and nine scouts in a minibus. They arrived in the early morning around 2.30am, and by 6.00am Mafeking was on the air. On this JOTA occasion, seventy-five stations were active through the Republic of South Africa, representing ninety-two scout groups.

Amateur radio is a classless hobby, and the words 'Monarchs themselves have not thought it derogatory to their dignity to patronise the art' have a special significance, since King George VI was a radio amateur and set-constructor. The Royal interest was continued to the 1971 JOTA, when among the total of 5,000 participants in the United Kingdom was HRH Prince Andrew. He helped to man station GB3RAC and was on watch for 2 hours during a Sunday communication session. At this JOTA 250 stations operated, and between them they worked 399 different scout stations in fifty-three countries.

Through the whole pageant of JOTAs, the World Bureau of Boy Scouts emphasises that (unlike many other aspects of amateur radio) it is not the number of contacts that count, but the contribution made towards world communication, understanding, and the cause of peace. Lt-Gen Lord Baden-Powell, OM, GCMG, founder of the scout movement, stressed this in person when he attended the 5th World Jamboree held in Holland in 1937. 'If you are friends', he said, 'you will not want to be in dispute. By cultivating these friendships you are preparing a way for solutions of international problems by discussions of a peaceful character. This will have a vital and flourishing effect throughout the world in the cause of peace, and so I pledge you all to do your utmost in establishing friendship among scouts of all nations.' The annual Jamboree-on-the-Air is a link by radio (and by slow-scan TV, too, nowadays) which Baden-Powell could never have visualised.

CHAPTER FIVE

GETTING A PERMIT

Ham listeners face few restrictions, although in countries where a duty has to be paid to the state for permission to operate a radio receiver, they should take care to read the regulations printed on the licence, which form part of the law of the land. In the United Kingdom, for example, no separate licence fee is now payable by listeners to the broadcast or the ham-radio bands, but restrictions still apply.

These are basically concerned with avoidance of interference, misuse of information received, and copyright. The current UK permit is issued under the Wireless Telegraphy Act 1949, and the regulations include such stipulations as:

> The apparatus shall be so maintained and used that it does not cause undue interference with any other wireless telegraphy . . . This licence does not authorise any infringement of copyright in the matter received . . . If any message other than a message for the receipt of which the use of the apparatus is authorised is unintentionally received, no person shall make known its contents, origin, destination or existence, or the fact of its receipt, to any person, other than a duly authorised officer of Her Majesty's Government, a person acting under the authority of the Minister of Posts and Telecommunications or a competent legal tribunal, and shall not reproduce in writing, copy, or make use of any such message or allow it to be reproduced in writing, copied or made use of.

GETTING A PERMIT

This somewhat pompous and cryptic copyright regulation, which every UK ham must observe, is primarily designed to prevent him from becoming an unofficial and irregular link in picking up and redistributing news bulletins or secret military information. In countries where there are no such provisions as those under the 1949 Wireless Telegraphy Act of Great Britain, there are invariably restrictions on the use hams may make of the information received off-air.

Operating an amateur transmitter is quite a different proposition from the legislator's viewpoint, because, of course, operation of unlicensed, unknown and pirate transmitters would quickly bring chaos to the ether. In the USA there are heavy penalties for operating unlicensed stations (not only ham stations, of course, for pirate broadcasting stations can be equal offenders), and currently these could bring a maximum of 2 years in gaol and a $10,000 fine. The same applies to offenders in Canada, but in the United Kingdom the maximum penalties were only a fine of £100 and a 3-months' sentence until a spate of pirate radio stations sponsored by illegal political and pop groups compelled the administration to increase them. Naturally, human nature being what it is, there is often a temptation for a tyro to operate a walkie-talkie without a permit or (since the cheaper imported dual-sets are fitted with crystal control for illegal transmission bands) on incorrect frequencies. From time to time newcomers in their over-enthusiasm modulate a simple oscillator with a microphone, and try to compete with the BBC or their local radio. On the whole, however, there are few offenders, and fewer still remain uncaught, for amateurs monitor their permitted bands very thoroughly, and so of course do Post Office detector vans. The untrained and unlicensed amateur could so easily jam an essential transmission, never knowing the harm he does.

When it comes to transmitters, whether it is a 'license' (US) or a 'licence' (UK), it is essential not only to have one and to renew it regularly, *but to possess the qualifications that type of permit requires*. These qualifications are perfectly logical. In the

GETTING A PERMIT

USA there are separate permits for novices, technicians, general class, advanced class, and amateur extra class, but in the United Kingdom the pattern is rather more complex.

Inquiries by would-be British amateurs should be made to The Home Office (Radio Regulatory Division), Waterloo Bridge House, Waterloo Road, London SE1 8UA.

There are five main groups of licence, as follows, and the figures given here show the fee payable on issue, and thereafter anually:

		£
1	Amateur (Sound) Licence A	3.00
2	Amateur (Sound) Licence B	3.00
3	Amateur Television Licence	3.00
4	Amateur (Sound Mobile) Licences A and B (supplemental to main licence)	1.50
5	Amateur (Sound Mobile) Licences E and F (the main licence). Available for those who wish to hold an Amateur (Sound Mobile) Licence only.	3.00

The A-class licence is the principal type of amateur transmitting licence, and to qualify you must be over 14 and a British subject. You must pass the Radio Amateur Examination and also the Post Office Morse Code test, details of which are given later. You will be allotted a call-sign. Three-letter calls are allocated in strict alphabetical sequence in these A-class licences, with the prefix G4 for England, GM4 for Scotland, GW4 for Wales, GI4 for Northern Ireland, GC4 for the Channel Islands, and GD4 for the Isle of Man.

The B-class licence is sometimes lightheartedly referred to as the 'chatters' charter', since it is for phone only, and does not authorise the use of Morse. Its conditions are broadly the same as those for the Amateur (Sound) Licence A, except of course that one does not have to take the Post Office Morse Code test. However, this licence is for hams to operate on a restricted band of frequencies, and does not authorise frequencies below 144MHz. Because of reluctance to take the Morse Code exami-

GETTING A PERMIT

nation, an increasing number of hams start with this licence, which naturally covers rigs of some complexity, owing to the use of higher frequencies than usual; in earlier days hams used to start on the 80m or 40m bands, then work their way down to 10m. With the new licences three-letter call-signs are issued alphabetically, with the prefix G8 for England, GM8 for Scotland, GW8 for Wales, and so on.

The Amateur (Television) Licence, like the B-class licence, necessitates passing the technical exam but not Morse. One cannot use frequencies lower than 432MHz. Call-signs are issued as with the 'G8s', but have the prefix G6, GM6 and so on, as appropriate. There is also a suffix /T, so that Amateur TV station ZZZ, for instance, has to give his call as G6ZZZ/T.

The Amateur (Sound Mobile) Licences A and B are, as their titles imply, only for mobiles. If you wish to qualify, you must be the holder of either a current Amateur (Sound) Licence A or B, and of course in the latter case you would be restricted to frequencies above 144MHz only. The call-sign is the same, with the addition of the suffix /M.

For Amateur (Sound Mobile) Licences E and F you must pass the technical exam and Morse Code test for Sound Mobile E, but the technical exam only for Sound Mobile F. Call-signs also carry the suffix /M.

Any reasonable ham would regard the conditions attaching to British amateur licences as fair and sensible. All of them are given in 'the small print' of the licence itself, which many hams may not bother to read. It should be emphasised, therefore, that a ham is licensed only to use his station for sending to and receiving from other licensed amateur stations 'as part of the self-training of the licensee in communication by wireless telegraphy'.

As to rag-chewing, he may send or receive only 'messages in plain language which are remarks about matters of a personal nature in which the licensee, or the person with whom he is communicating, has been directly concerned'. And he must not use signals in secret code or cipher. There is a relaxation for public-service emergency work, but the official wording, in its stilted style, refers to 'The use of the station, as part of the self-training of the licensee in communication by wireless telegraphy during disaster relief operations.' In plain English it means that the ham may take part with any other licensed amateurs in assisting disaster-relief operations conducted by the British Red Cross Society, the St John Ambulance Brigade, or the police.

The basic licence does not cover the use of a ham transmitter aboard ship (or within any dock or harbour), and the transmitter must be used by the licence-holder personally (although another licensed ham visiting the shack may temporarily operate the station in the presence and under direct supervision of the licensee). No message that is grossly offensive, indecent or obscene may be sent. There is a very small fringe public of amateurs who do express themselves strongly and carelessly on the air, but other hams quickly warn them. On the whole the spirit of ham working is one of an enjoyable fraternity, and very little is transmitted that is not already paralleled in drama and variety series on the professional broadcast channels.

On one point the Home Office is adamant. Hams must not try to emulate broadcasters. 'Messages shall not be broadcast to amateur stations in general', it says, 'but shall be sent only to (i) amateur stations with which communication is established separately and singly, or (ii) groups of particular amateur stations, provided that communication is first established separately and singly with each station in any such group.' This gives absolutely no legal standing to the various 'Free Scotland', 'Son of Radio Caroline' or the rest of the ether pirates, which have occasionally jammed the ether and risked endangering urgent service channels.

A great deal of genuine amateur experimentation is concerned with frequency stability. As we have seen, in 1924 amateurs were among the first to develop crystal control for stable prime oscillators. The British licence demands only that 'a satisfactory method of frequency-stabilisation shall be employed', and the ham must have equipment for frequency measurement. When code (as distinct from phone) is used, a ham must ensure that the risk of interference from key clicks is eliminated (not merely reduced to a minimum). 'At all times', the regulations insist, 'every precaution must be taken to avoid over-modulation . . .' (In simple terms over-modulation means that the speech or music is over-impressed on the CW carrier, so that 'splatter' results from excessive sideband formation; again, a ham, Richard Fusniak (G3TFX), was the first to invent an over-modulation suppressor for low-power AM.) The regulations go on to say that the licensee must 'keep the radiated energy within the narrowest possible frequency bands, having regard to the class of the emission in use. In particular, the radiation of harmonics and other spurious emissions shall be suppressed to such a level that they cause no undue interference. . . . Tests shall be made from time to time, and details shall be recorded in a log. . . .'

A ham's log is as important to him as a ship's log is to a sea captain. It has to be kept in one book, not loose-leaf, and must be an indelible record, not pencil. In the various columns, which may follow any layout that suits the ham, he must show the date, the time every call begins (including the tests), the call-signs of the stations from which messages addressed to the station are received, the times of establishing and ending communication, and the exact frequency (not merely the band) and class of emission in each case. All times must be in GMT for hams working in the UK.

Logs are published for amateurs. A typical ham reception log would have columns headed 'Date', 'Time (GMT)', 'Band MHz', 'Station Heard', 'Calling', 'Mode of Emission', 'Signals (RST)', 'Notes', and a final double QSL set of columns for showing whether a QSL card was sent or received. Mode of emission is noted in the code already mentioned – A1 (CW telegraphy), A3 (phone), F3 (FM or PM, phase-modulation), and so on.

It goes without saying that a ham interested only in scanning the globe for DX, and sending out cards, does not necessarily have a transmitter; but it is probably overlooked that in the UK

GETTING A PERMIT

and certain other countries, a ham transmitting station must be equipped with a receiver capable of receiving messages on the bands (and on the particular type of emission) covered by the licence. It is illegal to go on the air 'blind', without a means of receiving signals on the same band of frequencies.

As they have to be worked for through the technical examination and probably the Morse Code test, hams are naturally proud of their call-signs. Nor must they be overlooked. The callsign must be transmitted either in code or phone, for identification purposes, at the beginning and end of each period of sending, and whenever the frequency is changed. If the period of transmission is greater than 15 minutes, the call has to be repeated in the same manner at every further quarter of an hour. When keyed, the Morse must not be at a speed greater than twenty words per minute, no matter how expert you may become in code, nor how fast you send the rest of the message. The callsign has to be slow enough for all to read. When the call is given on phone, the 'Alfa, Bravo, Charlie, Yankee, Zulu' code must be used (see Appendix, p 155).

When listening to ham telephony, you probably find it informative to learn that the transmitter is using So-and-So's capacitors, or that the rig has been improved by rewiring with Thingummy's cables or a Doo-da Electronic Company's dish antennas. In the UK this is technically illegal, if not specifically tied just to a rig of individual components. Too frequent descriptions of trade names contravene Section 16 (Interpretation) (3) of the regulations, which state that the station must not be used for 'business, advertisement or propaganda ... or for the benefit or information of any social, political, religious or commercial organisation'.

When considering the licences available in the UK, it is advisable to recapitulate the abbreviations favoured by the Home Office.

Amplitude Modulation

- A1   Telegraphy by on-off keying, without a modulating audio frequency.
- A2   Telegraphy by on-off keying of an AM audio frequency, or by on-off keying of the modulated emission.
- A3   Telephony, double sideband.
- A3A   Telephony, single sideband, reduced carrier.
- A3H   Telephony, single sideband, full carrier.
- A3J   Telephony, single sideband, suppressed carrier.

Frequency Modulation

- F1   FM telegraphy, by frequency-shift keying, without the use of modulating audio frequency.
- F2   FM telegraphy by on-off keying of a frequency-modulated AF, or an on-off keying of an FM emission.
- F3   FM telephony

Pulse Modulation

- P1D   Telegraphy by on-off keying of a pulsed carrier without the use of a modulating AF.
- P2D   Telegraphy by on-off keying of a modulated AF by keying of a modulating AF or by keying of a modulated pulse carrier, the audio frequency modulating the amplitude of the pulses.
- P2E   Telegraphy by on-off keying of a modulating AF or a modulated pulse carrier, the AF modulating the width (duration) of the pulses.
- P3D   Telephony, amplitude-modulated pulses.
- P3E   Telephony, width (duration) modulated pulses.

The input power to any ham transmitter is assumed for licence purposes to be the total DC power input to the anode circuit of the tube or tubes, or (to use the official description) 'any other device energising the aerial'. This is sometimes an imprecise way of stating power, so for A3A (reduced-carrier single-sideband phone) and A3J (SSB, suppressed carrier) a figure is quoted for radio-frequency output peak envelope power. The initials 'erp' come up in connection with transmitter powers, and this is verbal shorthand for 'effective radiated power'. On A3A

and A3J the expression 'pep' indicates 'peak enveloped power'.
For the United Kingdom the table below is the schedule for the Amateur (Sound) Licence A, which of course necessitates taking both the technical and Morse Code tests.

Frequency (MHz)	Emission	Max DC Power (W)	RF PEP (W) A3A and A3J only
1·8–2·0	A1, A2, A3, A3A, A3H, A3J, F1, F2, F3	10	26 2/3
3·5–3·8			
7·0–7·10 14·0–14·35 21·0–21·45 28·0–29·7		150	400
70·025–70·7		50	133 1/3
144–145			
145–146		150	400
430–432	A1, A2, A3, F1, F2, F3	(10W erp)	
432–440 1,215–1,325 2,300–2,450 3,400–3,475 5,650–5,850 10,000–10,500 24,000–24,050 24,050–24,250	A1, A2, A3, A3A, A3H, A3J, F1, F2, F3	150	400
2,350–2,400 5,700–5,800 10,050–10,450	P1D, P2D, P2E, P3D, P3E	25W mean power	

There are certain restrictions in these bands in the UK. Radio Teleprinter (RRTY) must not be used in the 1·8–2·0MHz band; there are a number of specified spot aviation frequencies that

GETTING A PERMIT

have to be avoided in the 144–146MHz band; and there are coastal-area restrictions in the 430–432MHz band.

The British amateur is now fortunate, perhaps, that a B-class licence may be obtained without need to take the code examination. It is a relaxation that encourages the 'scientific ham', interested in VHF, UHF and microwave electronics, as distinct from the amateur who is thrilled by DX. A non-code licence restricts the ham to the frequencies of the 144MHz order. In the earlier days of radio it used to be a real achievement to work an Australian amateur (or a UK ham from Australia) on the 'twenty' (14MHz) band, and higher frequencies were considered for experts and laboratory workers only; but today it is a wide-open field for the B-class operator.

No matter what band is being worked, a ham must have a means of keeping the frequency stable, and of checking it. The quartz crystal is one answer. Until the development of the ammonia atomic clock (in which the high energy molecules of NH_3 vibrate in a cavity resonator), the world's most accurate timekeepers were controlled by the electric oscillations within quartz crystal. As a matter of interest, the NH_3 molecules vibrate at about 24 milliard (24 billion, in US terms) cycles per second, and the gaseous ammonia atomic clock is roughly ten times as accurate as the quartz crystal. In turn, the average quartz circuit has a time-stability 1,000 times as good as a high-precision pendulum clock, which may have an error at least 3 seconds a year.

For radio purposes the quartz crystal oscillator is the most convenient way of achieving electrically a periodically recurring process. Under the influence of an alternating voltage, quartz (and certain other crystals) can be set into electrical vibration. Opposite electric charges are produced on the various crystal faces, and these produce oscillations that may be applied to the grid of a tube, or to a transistor. Accuracy is to within 1 tenthousandth of a second over several months, depending to some extent upon the thickness (which also is a factor determining the frequency), and upon the radio-frequency crystal current (which may cause heating). In ham work several crystals can be made available to a circuit by manual switching, or by a solid-state diode used as a switch.

Transmitter circuits (and some circuits in receivers, such as the BFO) may use a crystal oscillator, working at one or more predetermined frequencies; or a variable-frequency (VFO) oscillator, the oscillations then being controlled by the (magnetic) inductance and the capacitance in the circuit. A little trimming capacitor ('variable condenser') may also be used across a crystal working at frequencies higher than 1MHz, to set the frequency precisely. The crystal current is usually of the order of 20–40mA (milliamps, or thousandths of an amp).

The embryo amateur has to know how to check his transmission, and this is one purpose of the technical examination. However, it is not only a matter of passing the examination, but of being able to show to Post Office, Home Office or other officials, when they call at the ham's shack, that he knows what he is about. Regulations lay down that,

> When his station is inspected by officers authorised by the Minister, the licensee will be expected to demonstrate that he can conform with the requirements: (a) To be able to verify that his transmissions are within the authorized band, and that no appreciable energy is radiated outside that band (b) To use a satisfactory method of frequency control, and (c) to ensure that his transmissions do not contain unwanted frequencies, such as harmonics.

The wording sounds somewhat like a threat, but in practice hams' homes are visited only when there has been a complaint, or when official monitoring shows there may be a flagrant disregard of the rules.

Essentially there are two kinds of wavemeter (frequency-standard meter), one using a crystal oscillator, and the other absorbing a tiny fraction of the transmitter output and showing its frequency on the calibrated dial. With a crystal-controlled transmitter, an absorption wavemeter is used to check that the desired harmonic of the crystal is selected. In practice it is not

GETTING A PERMIT

necessary for the crystal controlling the transmitter to oscillate at the fundamental frequency. It is usually a harmonic, and that has to be selected carefully to avoid error. It could be a 'second harmonic' (twice the fundamental), and the crystal-oscillator (CO) waveform is usually complex.

Home Office guidelines are as follows. The frequency-measuring equipment has to be sufficiently accurate: for example, operation in the centre of the 21·0–21·45MHz band would require frequency measurement to an accuracy of $\pm$ 1·0 per cent to be of any real value, whereas operation within, say, 10Kc/s of band-edge would require measurement to an accuracy of $\pm$ 0·05 per cent. When determining the proximity of transmission to band-edge, the possible bandspread due to modulation (on phone, for example) on the appropriate side of the carrier needs to be added to the frequency tolerance of the carrier.

There are various ways of using heterodyne wavemeters and crystal calibrators. In one technique the calibrator is coupled to the ham's own receiver, and the transmitter oscillator stage only operated, with all subsequent stages switched off. It is a simpler process in the USA, as all US amateur bands are multiples of 25kHz, so that a whole range of 25kHz 'marker' signals can be produced and heard in the monitor receiver along with the signal from the transmitter oscillator. To speed identification of the correct marker, some commercial frequency-testers also provide fundamental outputs of 50 and 100kHz. The US National Bureau of Standards provides a service regularly used by amateurs from two radio stations, WWV at Fort Collins, Colorado, and WWVH at Kauai, Hawaii. Full details are available in *NBS Frequency and Time Standards* (Publication 236), available for less than $1.00 from the Superintendent of Documents, US Government Printing Office, Washington DC 20402.

Standard radio signals of high accuracy are broadcast from 'Whiskey-whiskey Victor' on 2·5, 5, 10, 15, 20 and 25MHz, with a slightly reduced spectrum from the Hawaiian station 'Victor Hotel'. Both stations are controlled by atomic-clock-type regulators, and the accuracy is to $\pm$ parts in 10^{11}.

GETTING A PERMIT

These are continuous night-and-day transmissions. For 45 seconds at a time a series of audio tones is given on 440, 500 and 600Hz (c/s). On odd minutes a 600Hz tone is transmitted by WWV, and on even minutes a similar tone by WWVH. Voice announcements are also given – time checks in GMT, geophysical alerts known as 'Geoalerts', and a separate series of propagation reports and forecasts that deal, for example, with short-wave conditions between Washington DC, New York and London.

In the USA it is possible to dial (303) 499-7111 for these nation-wide transmissions of time and frequency. To distinguish transmissions from the two stations, WWV uses a man to announce 'voice segments', and WWVH a woman. Hams are strongly advised to get the current list of *NBS Time Standards* as a guide to standards available on the air.

Canada is not yet covered by a standard-frequency transmitter, but time checks of great accuracy are transmitted from CHU on several frequencies, including 3,330 and 7.335kHz.

Amateurs in Europe use the frequency-standard transmissions from Germany, Italy and Switzerland, but many follow the 'Mike-Sierra-Foxtrot' (MSF) Standards station at Rugby, a traditional centre of Post Office radio. Like its Washington counterpart, MSF runs continuously throughout the 24 hours, covering a range of frequencies from 16kH (about 20,000m) through the shortwave bands, including 2·5, 5 and 10MHz. Periodically the tone is cut and the call-sign MSF transmitted in Morse (*dah-dah, dit-dit-dit, dit-dit-DAH-dit*).

Provided he is equipped with a good commercial wavemeter, an amateur can fairly easily satisfy licence requirements. Crystal calibrators are oscillating meters whose output can be heterodyned with the transmitter frequency, the transmitter then being adjusted to zero point so that it is 'spot-on' with the meter; and the British authorities suggest that these calibrators, when used in conjunction with a general-coverage receiver, need only a 100Kc/s crystal for checking frequencies up to 4MHz. For higher frequencies the spacing between the 100Kc/s marker points is too small for accuracy, and a crystal of 500Kc/s or preferably 1MHz

GETTING A PERMIT

should be used. If the receiver covers only the ham bands, the bandspread scale will usually allow a 100Kc/s crystal to be used with sufficient accuracy throughout the HF bands.

With the simpler absorption meters, there is usually a jack-point in the unit so that the ham can listen through the meter. There is a pick-up coil, an HF semiconductor diode acting as a 'crystal detector', the all-important calibrated tuned circuit, and a micro-ammeter or milliammeter of low range, such as 0–1·0mA. The scale length and accuracy must be suitable, and the frequency coverage should extend up to the second (and preferably the third) harmonic of the radiated frequency, so that the presence of unwanted frequencies can be detected. For VHF and UHF transmitters probably the best technique is to measure the frequency of the fundamental oscillator as accurately as possible, and then to use an absorption meter to confirm that the wanted harmonic has been selected.

No technical quiz has to be taken in Great Britain if the ham proposes to confine himself to listening only. He can, if he is wise, register with the Radio Society of Great Britain and be given a BRS (British Receiving Station) number, or an ORS (Overseas Receiving Station) if outside the UK. There is no need to apply for a number if you have a full A or B transmitting licence, but with a normal five-digit BRS number you can get right into the hobby, and have a properly registered number to use with QSL cards. Associate RSGB members (under 18) currently have a four-digit number.

A ham should always have a preliminary period as an operator of a BRS. It will give him invaluable experience in understanding ham jargon and the intricate procedure of operating. There is nothing second-class about a BRS operator. The RSGB regards Receiving Members as just as important as those with a TX. When you think of it, there is little point in transmitting unless someone is listening! At the present time about half the RSGB registered list is of Receiving Members, and of course most of them go on to take the technical Radio Amateur examination.

This examination is conducted twice yearly (usually in May

and December) by the City and Guilds of London Institute, 76 Portland Place, London WC1N 4AA. Currently there is a fee of £2·50. Full details about the examination and specimen question papers over a period of 3 years may be obtained from City and Guilds of London Institute, 25p post free within the UK and overseas by surface mail.

It is a 'pass' examination, so you are not competing against other candidates. There is a single question paper of 3 hours' duration. Both the questions in Part I are compulsory, but Part II consists of eight questions, of which only six need be attempted.

At first the RAE syllabus may appear somewhat daunting. One of the most experienced amateur-radio pioneers, Pat Hawker, G3VA (author of several RSGB publications), when asked 'Are the examinations difficult to pass?' replied: 'Not really – provided you are genuinely interested. Of course, they require a certain amount of preparation and willingness to learn; but with care, and regular spare-time practice and training, there is no reason why anyone should not feel confident of obtaining a licence. Many do so within twelve months of starting from scratch.' Many candidates (but by no means all) study for the examination at a local Technical College or Evening Institute, or attend the special lectures often held by local radio clubs – so check to see if there is a suitable course in your locality. Many others follow correspondence courses, or are entirely self-taught. Some, of course, may still have a smattering of electronics drilled into them as radio or radar mechanics in the Services. Others will have an O-level groundwork.

It cannot be denied that some knowledge of mathematics is essential. There are a few basic formulae, but any good book on radio will show that 'school-level maths' are sufficient to cope with such things as Ohm's Law, resonant frequency, and the calcuation of power. If such things really do appal you, then get all the pleasure possible from registering as a BSR operator. The syllabus could dismay you at first sight, but further reflection may show that you could cope with it if properly taught.

GETTING A PERMIT

PART ONE

1 *Licensing Conditions*

Types of licences available and qualifications required of holders. Terms, provisions and limitations laid down by the Home Office in the Amateur (Sound) Licence A, including the notes appended and the schedule of classes of emissions and frequency bands.

2 *Transmitter Interference*

Frequency stability. Avoidance of harmonic radiation and of interference by shock excitation. Use of key-click filters and other means of preventing spurious emissions. Dangers of overmodulation. Devices for reducing interference with nearby radio and TV receivers. Need for audio bandwidth limitation. Frequency-checking equipment.

PART TWO

Elementary Electricity and Magnetism

Theory of electricity, conductors and insulators. Units. Ohm's Law. Resistors in series and parallel. Power. Permanent magnets and electromagnets and their use in radio work. Self- and mutual-inductance. Types of inductors used in receiving and transmitting circuits. Capacitance. Construction of various types of capacitors, and their arrangement in series or parallel.

Elementary AC Theory

Alternating current and voltages. AC theory incorporating circuits with inductance, capacitance and resistance. Impedance, resonance, coupled circuits, acceptor and rejector circuits. Transformers.

Thermionic Valves and Semiconductors

Characteristics of semiconductor diodes, transistors, thermionic diodes, triodes and multi-electrode tubes. Use of semiconductor devices and tubes as oscillators, amplifiers, detectors and frequency-changers. Distortion. Harmonics, Push-pull. Power rectification. Stabilisation and smoothing. Typical power packs for low-power transmitters and receivers.

Radio Receivers

Typical receivers. Principles and operation of CW reception. Reception of SSB and FM signals, in outline.

Low-power transmitters

Oscillator circuits. Use of quartz crystals to control oscillators. Frequency multipliers and power amplifiers. Methods of keying transmitters. Modulation. Types of emission in current use, including SSB and FM.

Propagation

Nature and propagation of radio waves. Ionospheric and tropospheric conditions, and their effect on propagation. Relation between wavelength, frequency and velocity of propagation.

Aerials

Common types of receiving and transmitting aerials. Transmission lines. Directional systems. Aerial couplings to lines and transmitters.

Measurements

Measurement of frequency. Operation of simple frequency meters, including crystal-controlled types. Use of verniers and other interpolation methods. Dummy loads and their use for lining up transmitters. Measurement of the power input to the final stages of a transmitter. Measurement of current and voltage at audio and radio frequencies. Use of cathode-ray oscilloscope for examination and measurement of waveform.

Probably along the way there are some elements of this syllabus that the reader of this book will regard as familiar, otherwise the book would not be in his hands! For a successful pass a year's study is likely to be necessary; but that year also gives one the opportunity to join the ARRL and/or RSGB, and in the latter case to gain some experience listening to the short-wave world.

The time factor is important. The Amateur Radio Certificate is issued free to British subjects who pass the RAE and the Morse Code test (after paying the examination fees, of course), and then you have exactly one year in which to apply for an A-class licence *after the date of the Morse Code test*. Allow more than a year to elapse and you will be required to take the Morse Code test again, but not the Radio Amateur Examination. The common sense behind this is that if an applicant passes the RAE, he presumably knows his subject, and is not likely to get 'rusty'; but unless Morse is used regularly it is apt to be something pushed into the back of the mind. So in a fit of enthusiasm do not take the Morse Code test first, and then swot up for the technical examination afterwards, because you might find yourself out of time, and forced to take the Morse Code test again.

How stiff are the exam questions a British applicant may face? Here are a few examples from recent papers.

> Explain how the following types of interference can be abated:
> (a) At the transmitter: (i) Harmonics, and (ii) Key clicks.
> (b) At the receiver: Image response
>
> Explain the meaning of self-inductance and mutual inductance, and define the unit of inductance.
>
> Explain why superheterodyne receivers are more selective. How do they differ from TRF receivers? Give an example of an image signal.
>
> State the relation between frequency and wavelength. State the ranges of amateur frequencies which are most suitable for (i) local, and (ii) distant transmissions.
>
> How would you detect the presence of standing-waves in a transmitter aerial feeder system.
>
> Draw the circuit diagram of a heterodyne wavemeter, and explain the use of this instrument for frequency-checking.

The general position of amateur licences in the USA has been briefly outlined in Chapter One. Essentially there are the following five categories, and all of them demand some knowledge of the Morse Code.

Novice

No previous experience is necessary, but the written examination must be taken for elementary radio theory and the essential amateur regulations. The term of this licence is 2 years, and it is not thereafter renewable. However, after 12 months without a licence, a ham may start all over again by retaking the examination. It is not intended that the novice licence should be 'for ever and a day', but is simply an opportunity for the newcomer to develop his skills, and get on the air legally. He has to be able to send and receive code at a rate of five words a minute.

Bands to which he is currently restricted are

$$3,700–3,750 \text{kHz}$$
$$7,100–7,150 \text{kHz}$$
$$21·1–21·2 \text{MHz}$$
$$28·1–28·2 \text{MHz}$$

He is restricted to A1 telegraphy – no phone, TV or any other emission. His maximum power is 75W. His transmitter does not necessarily have to be crystal-controlled, since an objective of the novice licence is also to give him experience with the VFO (variable-frequency oscillator). Unlike the British ham, he is not restricted to his shack, but may operate portable or mobile; and if he visits a friend with, say, a general class licence, he may accept an invitation to operate the transmitter – but only on Novice frequencies and with no more than 75W. He has to transmit his call along with that of his host.

Technician

The next stage in the range of US amateur licences is intended to give US citizens (or residential aliens) a wider experience of ham working, although as a matter of fact the permitted frequencies include those for radio control of boats, cars and airplanes. The technical examination is rather stiffer than that for novices, including a test that can be taken by mail under the personal supervision of a ham already holding a general class (or higher) permit, or a commercial radio operator. The Morse Code test is at five words per minute. This permit lasts for 5 years, and thereafter is renewable. It allows the technician to operate in the accepted bands of 50·1–54MHz, 145–148MHz, and above 220MHz.

General Class

No previous experience is necessary, although many holders

of this class began as novices. The necessary Morse Code speed is thirteen words per minute. The licence lasts for 5 years and is renewable. The applicant has to take the examination in the presence of an FCC representative, and such examinations are usually held twice annually; but if he lives more than 175 miles by 'airline' from one of the FCC field engineering offices (or is a US national living overseas, is in the armed forces or is physically handicapped) he may take it by mail. A conditional licence is issued to the successful candidate, but this carries all the privileges of a general class licence.

There are in fact eighty-four cities in the USA where these examinations may be taken, including the major field organisations in Boston, New York, Philadelphia, Baltimore, Norfolk, Atlanta, Miami, New Orleans, Houston, Dallas, Los Angeles, San Francisco, Portland, Seattle, Denver, St Paul, Kansas City, Chicago, Detroit, Buffalo, Honolulu, Anchorage, and Washington DC. Details of these, and of the many FCC sub-offices and other examination centres, are given in the ARRL guide, *The Radio Amateur's License Manual*, which is an essential book for every US amateur.

Advanced

This licence was known as Class-A in the 1930s, and has been 'Advanced' since 1951. It is an incentive to amateurs to work and study, and gives additional sub-bands, including 3,800–3,890kHz, 7,150–7,225, 14,200–14,275, 21,270–21,350, and 50,000–50,100kHz. If the examinee has already passed the general-class Morse Code test, he only has to face the tougher technical examination. The holder of a conditional licence (who therefore did not visit an FCC centre) must pass the advanced test under supervision of an officially appointed volunteer who holds a licence of the same or higher class.

Amateur Extra

This is the top US amateur licence, and applicants must have held a US or foreign licence above technician grade for at least a year. There is a code test of twenty words per minute, and a new examination has to be taken on advanced radio theory. There are extra frequency segments (listed by the ARRL) available as a privilege to the extra-class operators, and the District FCC Engineer-in-Charge issues a diploma certificate. The licence lasts for 5 years and is renewable.

As a guide to the technical level of the various US examinations, the following examples come from recent FCC papers.

Novice

Who is responsible for the proper operation of an amateur radio station? Where must the amateur radio *operator* and the amateur radio *station* licence be retained or displayed? What information must the station log contain? When using code, what is the meaning of CQ, DF, AR, SK? What are Q signals? What are the characteristics of a good A1 emission? What is a megahertz? How can AC be converted into DC? Draw the schematic diagram of an RF power amplifier circuit.

General Class

What types of one-way transmission by amateur radio stations are permitted, and which are prohibited? For how long must a station log be preserved? [This is not a trick question, though few seem to know the answer, which is 'For one year following the last entry'.] What are propagation factors influencing radio reception on the VHF amateur bands? How is voice information conveyed in FM and PM (phase-modulated) emissions? How can the peak-envelope power (PEP) from an RF amplifier be determined? What are some common RFI (radio-frequency interference) problems encountered by amateurs? Suggest some solutions. How do capacitors combine in series and parallel? What is impeddance-matching?

Advanced Class

What factors affect the state of ionisation of the atmosphere? Define maximum usable frequency. On what frequencies do SSB transmissions become more difficult? What effect would a reactive load have on an oscillator's output frequency? What is meant by percentage of modulation? Why is neutralisation important in amplifiers?

Extra Class
What frequency should a crystal oscillator circuit be tuned to for maximum stability? Discuss the merits of using choke-input versus capacitor-input filters in power supplies. What operating conditions are indicated by upward or downward fluctuations of a Class-A amplifier's plate current when a signal voltage is applied to the grid? How does the beat-frequency oscillator affect the tuning of an SSB signal? What is the image response of a receiver? How do nrn-type transistors differ from pnp-type? What is meant by 'end-effects' in an antenna? Explain the properties of a quarter-wave section of an RF transmission line.

A reader unfamiliar with electrical engineering may feel that this is 'Extra Class' stuff indeed, demanding at least an engineering degree. This is not so. Many would-be amateurs pass their novice stage with the aid of the ARRL's *How to Become a Radio Amateur*, and as far as their general class by studying those two 'Bibles', the RSGB's *Amateur Radio Techniques* and the ARRL's *The Radio Amateur's Handbook*.

If you do not belong to a circle of fellow-amateurs and students, you would be wise to work from a good elementary electrical textbook. Amateurs in the UK taking the City and Guilds examinations should certainly study RSGB publications, including *Radio Amateur's Examination Manual* and *Radio Amateurs' Revision Notes*.

CHAPTER SIX

DIDAH LANGUAGE

There are many amateurs who maintain that Samuel Morse ought never to have written down his code on paper, but that right from the very beginning in the nineteenth century it should have been expressed just as a series of sounds. Of course there are probably many students, working hard on their five words a minute for the US Novice License or the twelve words a minute for the British Post Office examination, who feel that the world would be a happier place had Morse not lived at all.

All the same, the Morse Code has proved essential for world communication. It is particularly effective for amateur DX working, and is essential for Moon-bounce and other erudite aspects of ham life and times. Everyone is now agreed that the only effective way to learn the code is to treat it as a series of sounds, and that it should not be memorised as 'dot dash' and 'dot-dot-dot', or visualised as · — and · · ·.

Nevertheless, to proceed in simple stages, here it is in printed form.

A	·—		H	····
B	—···		I	··
C	—·—·		J	·———
D	—··		K	—·—
E	·		L	·—··
F	··—·		M	——
G	——·		N	—·

DIDAH LANGUAGE

O	— — —	U	·· —
P	· — — ·	V	··· —
Q	— — · —	W	· — —
R	· — ·	X	— ·· —
S	···	Y	— · — —
T	—	Z	— — ··

Special Continental Morse Letters

ä	· — · —	à or á	· — — · —
é	·· — ··	ñ	— — · — —
ö	— — — ·	ü	·· — —
ch	— — — —		

Numerals

1	· — — — —	6	— ····
2	·· — — —	7	— — ···
3	··· — —	8	— — — ··
4	···· —	9	— — — — ·
5	·····	10	— — — — —

Punctuation and Procedure

(.)	Full stop	· — · — · —
(,)	Comma	— — ·· — —
(?)	Query	·· — — ··
(:)	Colon	— — — ···
(')	Apostrophe	· — — — — ·
(—)	Dash	— ···· —
(/)	Fraction bar	— ·· — ·
(()	Left bracket	— · — — ·
())	Right bracket	— · — — · —
Underline (transmitted at beginning and end of passage)		·· — — · —
Break sign (=)		— ··· —
Quotes		· — ·· — ·
Error		········
X (ending signal)		· — · — ·
Invitation to transmit (K)		— · —
Wait		· — ···
End of work		··· — · —
Starting signal		— · — · —

Until World War II most amateurs (and professional communications operators, too) learned the code the hard way; and then to improve the speed of signals personnel during the war devices were introduced with the philosophy of 'thinking in sound'. This had also been commercially introduced by techniques such as the Candler System. Walter H. Candler devoted his time from 1911 onwards to teaching students and operators how to regard telegraphing as primarily a mental process. The Candler Junior Code Course was being used by radio amateurs internationally in the 1940s. After instructing students in the techniques of keying, and of avoiding a rigid wrist, the course dealt with thinking in terms of code, word sounds, three-letter combination words and groups, and 'how to read code as you listen to someone talk'.

Later came systems of instruction on disc and tape. In the United Kingdom the RSGB issues LP gramophone records, and publishes a most useful book called *The Morse Code for Radio Amateurs*, which was compiled by a YL ham, Margaret Mills G3ACC. She was one of the first women to be commissioned during World War II as a signals officer in the Women's Royal Air Force, and the first woman to be granted an amateur transmitting licence after the war.

There are commercial courses, such as those organised by the British National Radio School of Jersey, the Tutorial Division of which is at 55 Russell Street, Reading, Berkshire. The BNRS course covers the whole RAE syllabus and the Morse Code test. In the USA the Series-500 Instructograph teaches the Code by means of special tapes, with a speed range from three to forty words per minute. This system is produced by the Instructograph Co, Box 5032 Grand Central, Dept E, Glendale, California 91201. There are systems such as those of the National Radio Institute, Washington DC 20016, which, like the British National Radio and Electronics course, covers the whole field of amateur radio. The NRI amateur training includes equipment (which students on the postal course keep) such as a novice-class 25W transmitter and a matching three-band superhet receiver, plus a transistorised code-practice oscillator with Morse key and headset. Naturally this ambitious way of learning about ham radio and the Code is a hundred times more costly than a simple Morse key and buzzer – but then this old-fashioned do-it-yourself buzzer method is scorned by every expert. Unless the buzzer outfit is fitted with a suppressor circuit, it can cause local TV interference. The volume cannot easily be regulated, and, more important, one has to learn with the one standard note of the buzzer; on the air of course the CW note can be varied from a shrill tone to a low bleep, and those who learn with a buzzer will need to refamiliarise themselves all over again.

Circuits for basic transistor oscillators are given in the ARRL's *The Radio Amateurs' Handbook*. The estimated cost of parts is less than $5.00. The British equivalent, published in Margaret Mills' RSGB book, is a circuit in which any transistor will oscillate.

It may be wondered why Morse did not work out a simple sequence such as one dot for A, two for B, three for C, and so on. Equally it might be asked why the keys of a typewriter have the curious 'Q W E R T' sequence, and are not placed in alphabetical order. Even brief consideration will show that the keys in fact are positioned so that those most frequently in use are either at the centre of the keyboard (such as 't', 'h' and 'b'), while the much-used 'e' and 'o' are placed at opposite ends of rows, so they can be hit by the fingers of left and right hands respectively. Morse was clever enough to use the simplest and briefest symbols for the letters most used – such as one dot for 'e', two dots for 'i' and one dash for 't' – leaving the longer symbols for letters such as J, Q and Z. Unfortunately the Morse Code came into use before American amateurs devised the Q code, for otherwise Samuel Morse might have made some changes to avoid Q being the rather long-winded 'dah dah di dah' it is.

Incidentally Boy Scouts of the Scout Association are taught that a young artist returning to Europe from the USA on the steamship *Sully* overheard passengers discussing an experiment they had seen in Paris, in which electromagnets were operated by current over very long lengths of wire. As a switch closed at one end, needles moved at the other! Before the voyage ended, this young man – whose name, of course, was Samuel Morse – had worked out his first dots-and-dashes code.

One has the feeling that all the mistakes that could be made in trying to memorise the code have already been made. If they have not led to failure, they have produced unnecessary mental anguish. An example is 'opposites'. We used to be told that a good way of memorising the Code was not 'dot dash' for A, 'dash dot-dot-dot' for B, and so on, but that A is 'dot dash' ($\cdot$ —) and N is its opposite 'dash dot' (— $\cdot$), that K is 'dash-dot-dash' (— $\cdot$ —) and R is 'dot-dash-dot' ($\cdot$ — $\cdot$) etc.

It probably ought to be accepted that if you simply cannot remember that S is three dots, T is a dash, and O three dashes, you certainly will not be able to go on to the Js, Qs and Ys, and perhaps you ought to take up philately, sailing or photography instead of ham radio. The truth, of course, is that millions of people can easily remember these simple groups, just as many millions more remember the outlines of shorthand or the syntax of a foreign language.

The ARRL stresses that you should consider the code simply as a way of conveying information, and not deliberately try to 'translate' words into dots and dashes. It says,

> Learning the code is as easy – or difficult – as learning to type. Think of it as a language of *sound*, never as a combination of dots and dashes. It is easy to 'speak' code equivalents using 'dit' and '*dah*' so that A would be 'di*dah*' (the 't' is dropped in such combinations). The sound 'di' should be staccato. A code character such as 5 should sound like a machine-gun burst: *didididit*. Take a few characters at a time. Learn them thoroughly in di*dah* language before going on to new ones. If someone who is familiar with code can be found to 'send' to you, either by whistling or by means of a buzzer or code oscillator, enlist his cooperation. Learn the code by listening to it. Don't think about speed to start; the first requirement is to learn the characters to a point where you can recognize each of them without hesitation.

DIDAH LANGUAGE

Margaret Mills does not go the whole way with this philosophy, and perhaps there are some people who cannot immediately stomach the fact that a comma in Morse is 'dah-dah-didi-dah-dah'.

Whether your mental make-up is such that it prefers . . . or dit-dit-dit for S is something you alone can determine, but the Margaret Mills method takes it in stages. Lesson One includes,

```
E   ·
I   ··
S   ···
H   ····
T   —
M   — —
O   — — —
```

Lesson Two continues to build up the dot-dash groups, thus:

```
A   · —
U   ·· —
V   ··· —
```

The converse is given in Lesson Three:

```
N   — ·
D   — ··
B   — ···
```

With these and subsequent lessons go groups of letters and words, so that a knowledge of the entire code is acquired by stages. Margaret Mills makes the wise suggestion that one should concentrate on letters that present difficulty, and automatically 'translate' into sounds the letters that have been learned as they are seen on hoardings and in newspaper headlines. The same goes for Q codes. If you are able to work with a colleague who is a radio amateur, it is possible to carry on quite a lengthy conversation on the lines of QRU (Have you anything for me?) . . . QRV (I am ready) . . . QRL (Are you busy?) . . .

DIDAH LANGUAGE

QRZ (Who is calling me?) . . . QRS (Send slower), and so on. In 'di*dah*' sound-language the code is as follows:

A	di-dah	N	dah-dit
B	dah di-di-dit	O	dah-dah-dah
C	dah-di dah-dit	P	di-dah dah-dit
D	dah di-dit	Q	dah-dah di-dah
E	dit	R	di-dah-dit
F	di-di dah-dit	S	di-di-dit
G	dah-dah-dit	T	dah
H	di-di-di-dit	U	di-di-dah
I	di-dit	V	di-di-di dah
J	di dah-dah-dah	W	di dah-dah
K	dah-di-dah	X	dah di-di-dah
L	di-dah di-dit	Y	dah-di dah-dah
M	dah-dah	Z	dah-dah di-dit
1	di dah-dah-dah-dah	6	dah di-di-di-dit
2	di-di dah-dah-dah	7	dah-dah di-di-dit
3	di-di-di dah-dah	8	dah-dah-dah di-dit
4	di-di-di-di dah	9	dah-dah-dah-dah dit
5	di-di-di-di-dit	0	dah-dah-dah-dah-dah
Query mark		di-di dah-dah di-dit	
Full stop		di-dah di-dah di-dah	
Comma		dah-dah di-di dah-dah	
Invitation to transmit (K)		dah-di-dah	
Preliminary call (CT)		dah-di-dah-di-dah	
Error (8 dots)		di-di-di-di-di-di-di-dit	
Wait (AS)		di-dah di-di-dit	
End of work (VA)		di-di-di-dah di-dah	

The ARRL *Learning the Radiotelegraph Code* booklet is similar to the RSGB book by Margaret Mills. In the USA a Code Proficiency Award is given by the League, which also transmits plain-language code practice material each morning and weekday evening, at speeds ranging from 5 to 35 wpm, from station W1AW; issues of *QST* give details of transmissions, some of which are given only from May through October. Throughout the world groups of amateurs transmit various code practice sessions on-air. One must take care not to get into the habit of slack sending: one 'dah' must always equal three 'dits', the space between letters should be equal to three 'dits', and the space between words must at least be equal to five 'dits'.

The Scout Association's *Communicator*, which instructs youngsters the world over for awards such as the Pursuit Badge, puts it this way:

> To a beginner, the letter S sent very slowly could be read as an S or as a series of Es, but three rapid dots cannot be mistaken. Even worse, a letter C sent slowly or badly can become TET, or KE or TR. As soon as you remember the sounds such as 'di-dah did it' for L, and 'Did it 'urt yer' for F, you recognise letters immediately, and do not have to stop and think how many dots and dashes they comprise. In time it is even possible to recognise whole words such as 'Dah diddley-dit-dit dit'. as the word 'the'. Once you have reached this stage you are well on the way to becoming a good operator. Try to memorise as much of the code as you can, even if it is only a few letters a day; and then whenever you have a few moments, try these out in your mind on any posters, advertisements or road-signs you can see. Even car number-plates will give you some practice.

Rhythm is important. A well-known amateur, Stan Bennett (G3HSC), has developed the Rhythm Morse course, which is based on ear training, and there is 'listening' drill right from the start, without bothering about printed dots and dashes. Bennett says,

> By receiving the letters quickly you can learn them as a rhythm. Once this rhythm has been mastered you have no longer any need to think. Morse is just another language. The idea is to train in the actual and unmistakable rhythms of the code, so that when you hear the 'tune' of a word, you at once understand it as well as you do the same word on the printed page.

This course has now been recorded on discs suitable for 33, 45 and 78rpm playing, using an LP stylus, and can be obtained from Belding & Bennett Ltd, 45 Green Lane, Purley, Surrey.

While the Morse Code test for the US amateur specifies five words per minute for novice and technician class, rising to thirteen for general/conditional (ten and fifteen words per minute respectively for the Canadian amateur), the test in the United Kingdom has more complex conditions attached.

In the *sending* tests a candidate must transmit thirty-six words (averaging five letters per word) in plain language in 3 minutes, without uncorrected error, not more than four corrections being permitted; and also ten five-figure groups in $1\frac{1}{2}$ minutes without uncorrected error, only two corrections being allowed. In the *receiving* tests the amateur has to receive thirty-six words (average of five letters each) in plain language in 3 minutes, and ten five-figure groups in $1\frac{1}{2}$ minutes. Each letter or figure incorrectly received counts as one error, while a word in which more than one letter is incorrectly received counts as two errors. More than four errors in plain language and more than two errors in the figure test result in failure. There are no punctuation marks or other symbols in any of these tests.

In Britain there are several schools at which guided tuition may be obtained by those who want to sit the Morse Code test and also the Radio Amateur Examination. Particulars may be obtained from the City and Guilds of London Institute, the Radio Society of Great Britain, or from the local education authority. Probably the most fun comes from learning in company with a handful of friends at a local amateur-radio club. The RSGB can provide addresses, and you can also get help from the main local Post Office, which keeps a check on licences. A good official source of UK information is the Radio Regulatory Division, Licensing Branch (Amateur and Special), Waterloo Bridge House, Waterloo Road, London SE1 8UA.

Ask almost any experienced amateur how he holds the key, and he will probably have to clasp his fingers around it to find out. The stance becomes automatic, and the aim is to avoid what US hams call the 'glass arm'. Margaret Mills says: 'A good keying "stance" is important, and a heavy Post Office type straight key is recommended. The key should be held lightly with the thumb below the knob and the first and second fingers resting on top. The action should be from the wrist.' One could say the same about holding a putter, or adopting a good golf stance generally, and in both activities one of the secrets is not to become tense.

CHAPTER SEVEN

ANTENNAS AND CHANNELS

Very few people are foolish enough to buy their first car and then announce: 'Well, now I'll take driving lessons.' It is putting the cart before the horse, in a manner of speaking. On the other hand, many over-impatient amateurs invest in an ambitious short-wave kit, a costly transceiver or a slow-scan TV outfit, and then for the first time start thinking about the type of antenna and the frequency bands to use it on.

When Marconi's 'S' signals first spanned the Atlantic, many people said it was a freak of nature, since, if electromagnetic waves could indeed be transmitted over long distances, there was nothing to prevent them flying off into space; and indeed this was very much more likely than that they would follow the curvature of the earth. Today we realise that the signals managed to cross the ocean, for one reason, because of the ionisation in the upper atmosphere, which is believed to be the result of ultra-violet radiation from the sun.

In a normal atom the negative charges on the electrons balance the positive charge on the nucleus, and if this balance is disturbed, the atom acquires a less negative charge (or, looked at the other way, becomes more positively charged), and is then said to be ionised with a positive ion. Nowadays ion exchange is an important industrial process, ranging from water-softening (hard-water calcium ions being replaced by sodium, through application of salt) to complex electrical and metallurgical processes.

Ionisation of a gas turns it into a partial conductor of electricity, since the moving ions form a passage for electric current. This applies also to the air. Test instruments in meteorological balloons, and experiments made with scientific satellites, have shown us that some 50–60 miles up from the surface lies a region of the upper atmosphere that is ionised in this way. It is not a fixed electronic envelope around us but is rather like the many layers of flaky pastry, drifting, moving and varying in density. We call it the ionosphere.

It plays an intriguing, infuriating and puzzling part in ham activities, because if the ionosphere did not exist, then waves, after being formed at the transmitting aerial, would go straight up and out into space. Only by reflection from the ionosphere do these waves return to the earth's surface. It will be realised at once that reflection, refraction and even absorption by the ionosphere is extremely variable. We are virtually at its mercy, since man cannot control sunspots or upper-atmosphere ionisation, so that all we can do is experiment. It is a never-ending facet of amateur and professional investigation. We use different types of antennas on different frequencies; we log the effects of a variable known as skip distance (areas where the sky wave returns to earth after 'missing out' on part of its journey); and we test directional aerial systems to learn about wave angles, fading, multi-hop propagation (conditions when a transmitted wave is reflected back from the ionosphere, then bounced back into it again from the surface), sunspot activity (this can cause ionospheric storms), and curious annoying ionosphere reflections resulting in what we call back-scatter, which is signal reflection from distances beyond the skip zone. If we knew all about the ionosphere, ham working would be more reliable but much less fun.

Experts draw up prediction maps to show with considerable accuracy the maximum usable frequency that will hold over any path on the earth's surface. Some such maps are published in

UK journals, including the *Short Wave Magazine* and *Practical Wireless*, while the full professional maps are issued by the US Government Printing Office in Washington, DC. Each set of four volumes of these maps costs approximately $10.00. The official title is *Telecommunications Research and Engineering Report, Ionospheric Predictions, OT-TRER*. Allied problems of 'troposcatter' are also the subject of continual experiment at the Marconi Research Laboratories at Great Baddow, Essex, England, and details are published in the private professional Marconi publication *Point to Point Communications*.

Scientists identify some ionospheric layers with letters, of which the E-layer is that nearest the surface, at a height of some 70 miles. The E-layer can exist only by gathering radiation from the sun, so its effect increases to local noon, then virtually disappears by sundown. Up at the F-layer, some 150–175 miles away, the air is thin, so that there is only a slow recombination of ions and electrons. Recently experimenters have been able to differentiate between the varying effects of two slices of this layer, so we now have an F_1 and an F_2 layer. There is a further complication for radio hams, since at various times of day a signal high in frequency may pass through the E-layer, but still be sent back again to one of the F-layers, so zigzagging its way round the earth. Amateurs have taken part in several experiments concerning 'sporadic-E', and while the mathematical and meteorological theory is complex, the results depend on hams patiently sitting at a Morse key, their Amplivox headsets on their ears.

Variable intense sporadic-E ionisation is not directly related to daylight or darkness, appears to be worse in equatorial regions, and can affect channels from 3·5 to 28MHz. Sporadic results apart, we can only say that some channels stand a better chance than others.

In the United Kingdom the top band is 160m (1·8MHz), and while this is not too susceptible to radio interference, there is nearly always heavy jamming from shipping and coast stations. During daylight hours Morse and phone are limited to some 75 miles, but in wintertime, when it is dark over most of the circuit, it is possible for UK receivers to pick up US ham stations. Because of the relatively lower power of the British amateur TX, the reverse is seldom true. This is not a DX band, but hams will appreciate getting a QSL card if stations are worked over more than about 500 miles at night-time.

The 80m (3·5MHz) band is also shared with shipping and some other commercial transmitters, and the best ham results come usually during the night or early morning.

The 40m (7MHz) band is rather like wine. There are good years and bad. Unlike wine, however, it appears to be affected by 11-year cycles of sunspots, as in 1958 and 1969. But a far worse enemy to the '7 MHz' than the ionosphere is the bevy of broadcasting stations that poach on it in open defiance of international regulations.

The 20m (14MHz) band is also affected by what are called 'sunspot minimum' years, when reception is worst. This channel in the European area of the world is about the best for DX, especially at dawn and dusk, but it may fade completely during winter nights. The companion 21MHz (15m) band has somewhat similar characteristics and is sunspot sensitive. In the spring and autumn the '21' is busy with American and even Japanese amateurs, and with suitable ionospheric conditions this can be sheer magic. Amateurs led the way in discovering that on 15m and 20m there can be a notable difference at various times of the day and year between east-west paths (as between the USA and Britain) and the north-south channels (as from Britain to South America).

The 15m and 10m (28MHz) bands are also variable, but DX is impossible at night on '10', except under freak electrical conditions. It is essential to study sunspot predictions and to make the most of evenings and early mornings in winter. Again, there are variations between north-south and east-west paths, British amateurs often finding that, although the ether seems 'dead' to their US friends, signals stream in from South Africa and Latin America.

ANTENNAS AND CHANNELS

In quite a different category are the 70 and 144MHz (4m and 2m) bands, which are the beginning of UHF technical interest and are not really suitable for globe-spanning DX. Here we are in the world of sophisticated solid-state devices, integrated circuits and highly directional aerials, with the emphasis on scientific electronics rather than communication. These signals are virtually not reflected by the ionospheric envelope, but are to some degree affected by sunspots and other electronic conditions in the upper atmosphere. We used to believe that the 2m band was limited to a range of up to 50 miles only, and radio clubs and scout radio groups did valuable pioneer work in what were thought to be just short hops, using fascinating equipment. Then QSL cards began coming in from other UHF ham workers, proving that distances up to several hundreds of miles could sometimes be covered.

Take care not to confuse the two '70s'. There is the 70MHz (4m) band and also the 70cm band. The latter, a frequency of 425MHz, rightly belongs in the UHF class and demands specialised equipment totally different from that in the average shack with a communications outfit. There are also bands above 1,215MHz that are also in what might be termed the ham laboratory category.

As an incentive to amateurs, the RSGB has introduced targets such as the British Commonwealth Radio Reception Award (BCRRA), which requires cards from at least fifty call areas in the Commonwealth. There is also the companion DX Listeners' Century Award, which, as the title implies, is for amateurs (registered listeners, not necessarily those owning transmitters) who chalk up a proved reception of stations from 100 different countries. Further encouragement for registered listeners, to participate in the very directional UHF bands of 70MHz and upwards, is offered by the RSGB awards for outstanding reception results in the 70, 144 and 420MHz bands; and these are known by the rather nostalgic-sounding title of 'Four Metres and Down'.

The electromagnetic energy of a transmitter, whether on 2m

or 160m, has to be fed into the ether by an antenna radiator system, which may be a simple long-wire 'aerial' or a highly-directional dish. Power is fed to the antenna by a feeder system, and at the receiving end there may be a similar efficient collector and feeder to the first tuned circuit in the shack.

Coupling transmitters and receivers to the line brings in a branch of electro-physics involving the mathematics of waveforms to such a degree that many amateurs disregard the theory and stick to rule-of-thumb. However, there are some simple analogies.

In the open air, when we want to shout some distance, we cup our hands to our mouths to form a directional trumpet, and if we are at the receiving end, we cup our hand to our ear. As we saw in Chapter One, radio waves are akin to light waves, and the fastest that an electromagnetic field can travel is 186,000 miles per second (300,000,000m per second). At the highest frequencies radio waves behave much as do light waves, though they can be more readily directed and reflected. Hitch a long single-wire antenna to any good communications receiver and you will hear something. In pure physics the antenna system should match the impedance of the receiver's input. In practice there is bound to be some electrical harmonic of the long single-wire aerial that matches the tuning of the receiver, but of course it may be that, were we to use a highly directional dish or system of rods with a correctly matched feeder system to the receiver, we should receive a much stronger signal or be able to attain greater DX in that particular direction on that particular frequency.

One says 'may', because a single-wire antenna some 30ft above ground, erected in an area reasonably away from walls and trees and hills, can give better results than an elaborate directional array that is incorrectly matched. Where space permits, amateurs sometimes arrange two or more aerials of different lengths and facing in different directions; they then connect a low-loss switch (low-loss from a high-frequency current viewpoint) or wander-lead-and-plug arrangement from the communications receiver to the best antenna for the time of day/night or DX direction.

ANTENNAS AND CHANNELS

Antenna design becomes more critical on the higher frequencies: a good single-wire aerial may be quite effective on, say, 1·8MHz, but from 10m down it is necessary to relate the size and proportions of the aerial to the size of the wavelength being handled – just as a tuning-fork resonates best when at the same frequency as that of an adjacent fork, or at least a simple harmonic of it. For this reason we use wires or rods arranged to match a half or quarter wavelength, and the double-wire (or coaxial) feeder is also matched to produce a 'flat' line over most of the amateur band in the particular frequency range.

We may have quite an elaborate layout of rods, usually three to six 'elements', in the form devised by the Japanese physicist Yagi, or one of the many other forms, such as the inverted-V, the multi-band trap or the end-fed Hertz. Numerous books have been written about antenna and feeder theory, and many hams regard it as one of the most fascinating branches of the hobby.

The complex antenna arrays one sees in the backyards and on the roofs of many amateurs may be a mute tribute to two US hams, W2NLY and W6QKI, who began experimenting 20 years ago with a thirteen-element Yagi formed of aluminium rods ⅛in in diameter and standing 24ft high. This is now a standard form for 144MHz working, and shows that in the shack below a truly serious-minded amateur is at work. There is no secret about this complex layout, for the two hams concerned published their results in *QST* for January 1956. Two years later what is now termed a broad-band dipole came into use for the 80m and 40m bands. Unlike the Yagi, it does not have a 'name' attached, but in fact it was developed by the staff of MIT (Massachusetts Institute of Technology) for radar experimentation, and at MIT it was known as the 'double-bazooka'. An amateur, W8TV, adapted it for use on 3·5, 4MHz and similar bands, and published the results in *QST* for July 1968.

Half-wave dipoles formed of thin-section rods tend to droop after a winter of bad weather, as do H-form TV aerials, but there is an easy way to overcome the problem. A simple half-wave dipole comprises an antenna of two wires or rods 'fed' in the centre by a coaxial lead, the total length of the aerial being approximately equal to half the operating wavelength. The best way to prevent drooping is to stretch it out between the top of a mast or tree and an anchor stake in the ground. The antenna slopes outwards, rather like a tent guy-rope; although at first glance it appears to be a single wire, it is in fact in halves. At the centre are two insulators, with the coaxial feeder (or two parallel wires) running out to them, so that both the halves of this sloping dipole are fed. Although the wire may slope at an angle of 45 degrees, the direction of maximum radiation is parallel to the earth's surface and not, as might first be supposed, pointing up to the sky.

Naturally there has to be some way of measuring the efficiency of an antenna system. We assess this in terms of 'decibels (dB) gain', using a simple dipole as a basis. As the dB is a logarithmic term, obviously the gain does not come in arithmetic progression. For instance, if the ham transmitter is working at 10W and we use a better aerial with a 3dB gain, this is equal to doubling the transmitter power; in the same way a 6dB antenna gain equals about four times the power, making the signal strength of a 10W station seem like that of a 40W one when registered by a distant receiver by means of an S-meter.

The bulb in a car's headlamp has to be placed at a precise point with regard to the focal distance of the polished reflector to send the maximum light out in a parallel beam; and although one cannot see radio waves, much the same conditions apply. The shape of even a simple dipole, or the small 'reflector' rods which are such a familiar part of a toast-rack-type of domestic TV aerial, is not a haphazard matter. The position of reflectors can be critical.

The subject is studied in depth in the 'Transmission Lines and Antennas' section of the ARRL's *The Radio Amateur's Handbook*, while the RSGB's *A Guide to Amateur Radio* (by G3VA) deals with the translation of antenna theory into ham practice. What the amateur has to keep in mind is that to get maximum efficiency – certainly with a transmitting aerial array – there has to be a completely matched system from the tuned circuit to what is called (according to one's country of origin) the 'antenna and ground' or 'aerial and earth'.

There are two sides to this circuit, which is not a DC circuit with a positive and negative flow but an AC circuit along which the waveform is travelling. This is on both sides of the twin feeder or coaxial, simultaneously. Although electronic flow is rapid, it takes a certain time; the wave does not start from the transmitter and travel along the feeder to the aerial instantly.

Picture a parallel pair of feeders, with a waveform (a varying applied voltage) in at one end and out at an infinite distance.

In 1 microsecond (one-millionth of a second) this wave will travel 300m, approximately 1,000ft. What happens to it? Well, as the two wires are parallel (or even enclosed one within the other in a coaxial cable), there is a certain amount of capacitance between them; and as they are made of metal and have an appreciable physical length, they also exhibit some of the properties of inductance – much as if each of the two wires was an infinitely long series of coils.

So, as the AC current waveform sees it, it is being poured into a whole network of capacitance and inductance, of infinite length. By changing the characteristics of the feeder we can make the waveform 'see' any future we wish. What is most important is that we can change the characteristic impedance or 'surge impedance', and, as far as the electron movement is concerned, we can then match the feeder to the tuned circuit of the transmitter, and to the antenna.

Thus, if we start a 10MHz wave off at the beginning of the feeder, and it is travelling with the velocity of light, in 0·1 microsecond it will have moved 30m along the lines. We can calculate and even measure the peaks and troughs in its passage. For example, when a transmitter is set up, an RF ammeter is used at various sections of the line to check on balance.

It is simplest to regard a simple dipole antenna as really the open end of the twin feeder, and the same theory applies even when a two- or three-element rotary beam aerial is used. The entire antenna array is rotated to any desired compass direction to obtain the maximum directivity and dB gain. If an amateur in the UK likes to concentrate on one direction – for example, on communication with American amateurs – he can used a fixed beam array.

Each shack is certain to possess a Great Circle map, and also a calculated antenna polar diagram. The first (available in the UK from the RSGB) shows how the world looks to radio waves emanating from a UK aerial. Such a map is essential when one is planning an antenna array, because radio signals inevitably travel the shortest routes on the globe. The polar diagram, which applies to each individual aerial, will probably have lobes showing zones of maximum and minimum radiation. Because of local conditions, aerials do not radiate (or receive) equally all around the 360°, and the aim of the experimenter is to get the maximum-lobe areas in line of sight between transmitter and receiver.

Again, the amateur has to consider how his transmitted electromagnetic waves are polarised (confined to definite planes). In theory this is determined by the position of the antenna radiator with respect to earth, a vertical aerial radiating vertically polarised waves. If the antenna is horizontal, it will tend to produce horizontally polarised waves, but there is a problem here because, for the effect to be at maximum, the horizontal antenna has to be at least one half-wavelength above ground.

These factors and many others call for rigid antenna construction, as well as observance of safety regulations. Feeders and aerials must not be placed where they could constitute a fire or safety hazard for the ham himself or for other people, or where the lines might blow down and come into contact with power or phone cables. Insulated guy wires must always be provided if there is a chance that a tower or rotary array may collapse in a high wind.

Most antenna construction of the timber-mast type is basic carpentry, and masts can also be made up from metal or wooden scaffold poles, or even from the PVC tubing used for rainwater pipes and conduits. Because of the greater space generally avail-

ANTENNAS AND CHANNELS

able in the American countryside, US hams probably have a slight edge on those in other countries when it comes to designing and constructing elaborate outdoor arrays.

Designs are published frequently by Short Wave Magazine Ltd, which also issues DX Zone and other Great Circle maps. Specialist suppliers of antenna equipment include Cohen TV Aerials Co Ltd, 645 London Road, Westcliff-on-Sea, Essex, England; Mosley Electronics, 40 Valley Road, New Costessey, Norwich, Norfolk, NOR 26K; Partridge Electronics Ltd (G3CED/G3VFA), Broadstairs, Kent; Strumech Engineering Co Ltd, Coppice Side, Brownhills, Walsall, Staffs; Kirk Electronics, 400 Town Street, East Haddam, CT 06423; Skylane Products, 406 Bon Air Avenue, Temple Terrace, FL 33617. Suppliers in the USA include Hy-Gain Antennas, Hy-Gain Electronics Corporation, 3601 North East Highway Six, Lincoln, Nebraska 63507; Mosley Electronics Inc, 4610 Lindberg Blvd, Bridgeton, Mo 63042; and Telrex Beam Antennas, Telrex Laboratories, Asbury Park, NJ 07712.

CHAPTER EIGHT

RIGS AND SHACKS

With a broadcast-type receiver capable of tuning down to 16om (1·8MHz), one can eavesdrop on Top Band ham-working, and strain the ears to pick up amateur code and Morse amid the cacophony of coastal stations. This spot will be found right at the bottom end of the medium-wave scale on some sets. Less than an hour of this experience is sufficient to demonstrate why the ordinary home receiver is not very satisfactory as part of a ham outfit.

Obviously, even if it covers the top band, such a receiver is unlikely to tune to all the higher-frequency amateur bands, and unless the UK amateur can tune to 20m and below, he has very little chance of genuine DX working. After the first initial thrill of picking up a French or German amateur, one realises that if the volume control is turned up in an attempt to hear even fainter (though not necessarily more distant) stations, the background tube or transistor hiss drowns the signal. Broadcast receivers are designed to receive programmes from a handful of stations with ample available volume and a minimum of distortion. The enthusiastic ham has quite different requirements. If he is successful in receiving a faint and distant station on a broadcast receiver, he will be disappointed to find that the signal fades after possibly only a few seconds. Having progressed thus far in this book, he may begin advancing his own theories about sunspot fading, the ionosphere and DX skip-distances; but in fact the fading is more likely to be due to electrical or mechanical instability within the set itself. Stations drift. Tuning and amplifier circuits within the receiver may be susceptible to heating, or to extremely small mechanical vibration. Even jotting down a call-sign on a nearby pad is enough to put some non-specialist receivers off tune.

As tuning for the louder broadcasting stations is not critical, some broadcast receivers have coarse rotary tuning arrangements, or possibly only press-button tuning. Such tuning is quite unsatisfactory for ether-searching on short waves, a procedure that demands vernier-like tuning controls with almost microscopic accuracy and refinement. As relatively close broadcasting stations have widely spaced frequencies, by international agreement, a high degree of circuit selectivity is not called for in the average domestic receiver; on short waves, however, the serious amateur must be able to tune between the many stations that almost 'sit' on identical frequencies. Numerous ham transmitters are themselves not as stable as major broadcasting stations, so at the receiving end the amateur has to hunt for wandering transmissions with the dedication of a Sherlock Holmes.

Most receivers these days, for TV and sound broadcasting, are of the superheterodyne type: that is to say, the signals are received at Stage 1 of the circuit and mingled (heterodyned) with oscillations of a nearby frequency, so that the outcome is a third resultant frequency dependent upon the intermediate frequency (IF). In many broadcast receivers this IF is set around 470kHz, and subsequent stages of amplification are handled at this new frequency (or, in some receivers, heterodyned to yet another different IF), before finally being detected, when the audio element is extracted and AF-amplified.

It is an excellent arrangement, but there can be disadvantages that militate against a simple broadcast receiver being used for serious communications work. In tubed sets the first stage is usually a combined mixer and local oscillator, and in the 'double superhet' layout there is a second oscillator. At the first stage the signals are converted to a high IF of about 2·0MHz, to give good image rejection. In the second stage the IF is heterodyned down to a much lower IF – perhaps the conventional 470kHz or even as low as 50kHz. The same principle is used with transistorised (solid-state) receivers. As can be imagined, spurious signals may arise from the single or double conversion, and from the operation of one or two inbuilt oscillators. A communications receiver can be planned to prevent this happening, but it is design error of this nature that causes spurious signals and oscillations (echoes and whistles) to drown the very faint signals the amateur is straining to hear.

In Chapter Two we saw that code can be transmitted on CW, and this must be heterodyned with yet another local oscillator at the receiving end to produce a difference frequency that is within audible range. By making slight changes in this internal oscillator – the beat-frequency oscillator – the operator can adjust the note to obtain the best conditions for code-reading. Without the BFO the Morse comes through as merely a series of clicks, or possibly it cannot be heard at all. A somewhat similar local oscillator is necessary to make possible the reception of phone systems such as suppressed-carrier telephony. These BFOs are naturally not required in broadcast receivers.

It probably goes without saying that an amateur's receiver must be sensitive. This does not mean simply adding extra valve or transistor amplification stages, because then turning up the gain may only increase the background noise. What is needed is circuitry with a good signal-to-noise ratio, so that even at maximum gain-control setting a faint signal can be heard over the mush. In the early days of radio most of this unwanted noise came simply from the mains hum, or from such defective components as carbon-track potentiometers and volume controls; but nowadays, with high-magnification circuitry, there are new sources of noise.

Beginners are obviously not going to design their ham receivers and transmitters, though it is possible to start as an amateur by building one of the many fine outfits provided by a Heathkit. The

Heath organisation is part of the international Schlumberger group, and Heathkits are manufactured by Heath (Glos) Ltd, Bristol Road, Gloucester, England, GL2 6EE, and by the Heath Company, Benton Harbor, Michigan 49022, USA. In any such kit the designing has been done, but the inquiring amateur needs to be informed of specification details that should apply to ham equipment, and the reasons why a standard broadcast specification is inappropriate.

Noise of any sort is an arch enemy of concentrated ham listening. In the early days of transistor development it was generally true that tubed receivers properly designed for communication work were less prone to background noise. Today transistor circuitry can be used even on UHF. Silicon solid-state devices are better for many purposes than germanium: they are more temperature-stable but more expensive, and although certain types are apt to produce more background noise than germanium, the general level is now at least as good as with tubed circuits. All transistor circuits draw low power from low-voltage circuits, so that they can easily be fed from dry batteries or even from a car battery. A 9V supply from a small PP3-type battery is adequate for many solid-state ham receivers, so that the total cost of operation is less than for valve equipment. A. L. Mynett, G3HBW, designed a very straightforward superhet receiver, which can be built at home for about £5 (around $10.00), and this has now been officially adopted by the Radio Society of Great Britain as a means of introducing newcomers to home construction. It has become known as the 'RSGB Transistor Four'; details are available to RSGB members, and are published in the RSGB's *A Guide to Amateur Radio*. Only simple metalworking is required, plus the ability to read a circuit and wiring-board layout, and to handle a small soldering iron carefully enough to ensure that excess heat does not damage the little OC170 and OC71 transistors.

Sophisticated communications receivers are sometimes fitted with an integral automatic noise-limiter (ANL), which removes some of the worst electrical peaks produced by what is known as 'site noise' – interference from power mains, neon signs, rail and road traffic, elevator motors and other imperfectly silenced sources of trouble. Obviously site noise is worse in cities, but some unfortunate ham shacks are cited in open country where the otherwise harmonious electrical conditions are completely nullified by static and hum from overhead power cables. It is equally obvious that if a filter ANL circuit is used to cut down man-made interference, there will be a limit to the workable sensitivity of the receiver. It is just a natural limit imposed on the amateur by his location.

If the shack is a room in the house, try to ensure that the receiving equipment is not being operated immediately under or over such noisy electrical gear as thermostats, motors, fluorescent strip lights, time switches or thermostat-controlled cooker switches. Imperfect contact at radiator and kettle power-points will provide a puzzling cacophony of crackling, even though the domestic equipment appears to work satisfactorily. Check trouble spots by feeling the plugs and wall sockets, when the power has just been turned off at the main. If there is local heat, there will be radio noise.

In a superhet receiver the frequency-changer valve is apt to produce noise, and that is why some good communications receivers have one or two good stages of RF amplification before the frequency-changer stage. Many broadcast receivers have no preliminary RF stage at all. A simpler 'straight' (tuned-radio-frequency, TRF) receiver is not subject to this inbuilt source of noise, though it is usually not very selective. In the crowded ham channels a receiver must offer almost knife-edge tuning, to cut out unwanted stronger signals while you are concentrating on the weaker DX signal.

The type of receiver that is pleasant to operate on HF and VHF has what is termed a 'bandspread' device, so that stations do not appear crowded together on the dial. A mechanical system may be used, with a vernier-like slow-motion tuning mechanism that must be completely free from lost motion (backlash). Alternatively the bandspreading may be achieved by having one or more very small variable capacitors, which are operated separately after the receiver has been rough-tuned on the main vernier dial; shifting the bandspread capacitor then will change the tuning by only a few kilocycles, one complete revolution of the bandspread dial being equal to perhaps only 1 degree on the main dial.

Stability is a feature somewhat easier to achieve in solid-state receivers. Completely tubed and hybrid layouts (mixed tubes and transistors) may take a quarter of an hour to warm and settle to a stable working. As in some professional television equipment (even of the solid-state variety), a cooling fan may be provided to help maintain a stable temperature; but the amateur philosophy generally is that the mechanical movement of the fan, its possible current variation and of course its noise, make the attainment of receiver temperature stability by good design a better proposition. Even a very small degree of temperature change (generally this means warm-up after 10 minutes or so) may affect inductance and capacitance values. In a well-designed receiver the essential RF screening does not 'build in' components that are temperature-susceptible or heat-producing. Valves are screened, but placed above most other components, so that heat escapes.

The chassis should be heavy. There should be no long connecting leads that may vibrate and so vary the RF tuning. If bandspreading is effected with a vernier tuning mechanism driven by a nylon-thread belt from one dial arbor (spindle) to another, the spring-tensioning device should not allow slip or judder. One general cause of instability in a communications receiver is insufficient regulation of the high-tension (HT) supply. This applies particularly to HT circuitry feeding IF and BFO oscillators, which of course are voltage-sensitive. There is fortunately a simple cure for HT instability, and this is to place a neon voltage-regulator tube in series with a small resistance, which allows a current of some 5mA to pass through the neon; the oscillator is fed directly across the neon, with a capacitor across the supply. Many receiver-construction manuals give details of this simple arrangement.

If you listen to 'ham-natter' on the short-wave bands, you will hear frequent references to 'lining up the IF' and to 'cutting out the crystal'. In a superhet receiver the incoming signal is translated to a different frequency and put through the subsequent stage or stages at this new frequency. Maximum RF amplification obviously depends on each IF stage being matched to the next. We need accurate tracking of the signal frequency and oscillator, so that the outcome is the constant IF figure. Therefore periodic tests are made with a signal from a test oscillator, and each IF stage is mechanically tuned to the correct point in turn. For receiver alignment in general, it helps if the frequency coverage of each individual waveband is kept to a minimum. Proud possessors of communications receivers covering an unnecessarily wide range might not be so satisfied if they realised how efficiency tails off at the end of each band. It is better to have more separate circuits or even plug-in coils and tuning assemblies. 'Rolls-Royce' quality communications receivers introduced to ham circles soon after the war – sets such as the National HRO – had up to nine coil packs. A slightly different band-change arrangement was adopted with the RCA AR88, another fine receiver (still available through the Used Equipment columns of journals such as *Practical Wireless* and the *Short Wave Magazine*).

Receivers such as the Hammarlund SP-600JX, the wartime R1155, the Eddystone EC-10 and a range of Collins equipment today have collectors' value among knowledgeable amateurs. These are tubed receivers, of course, but amateurs seek them out, and they are not too difficult to modify. The multi-band objections do not apply to more modern outfits such as the Racal RA-117 professional communications receiver, which covers 0·5–30MHz with the Wadley loop continuous tuning. One reason why old receivers are so much in demand is that they can easily be preceded with a crystal-controlled VHF converter.

On-air reference to 'cutting out the crystal' applies to superhet receivers, which receive CW through a crystal filter. A crystal can be 'cut' (shaped by grinding and by chemical etching) to operate at a fixed frequency with great stability. It can then be incorporated in a tuned circuit in conjunction with the RF/IF chain of a receiver, to form a rejector of unwanted frequencies. It improves the signal-noise ratio, and although in theory there could be a very small circuit loss, its main effect is to improve receiver performance enormously. The crystal filter has a variable capacitor in series, and this is known as a 'phasing' control. Clumsy operation of this control for BFO adjustment can indeed result in reduced signal strength, and that is why one sometimes hears amateur transmitters complaining that they 'need to cut the crystal' for reception of a faint signal. Technical journals from time to time publish features dealing with the correct operation of crystal filters, and it is certainly no unnecessary refinement on a good communications set.

Receiver techniques are changing, and sometimes improving, all the time. In the USA there is now a wide range of receivers with very Anglo-Saxon or even Space Age names, but probably the only component manufactured in the United States is the name label replacing a Japanese one. Many are beautiful examples of mass-production engineering, but because they are all aimed at the same market, they tend to be uniform in circuitry. Keen amateurs are willing to go out on a limb and experiment with such developments as the synchrodyne or 'DC' (direct-conversion) layout. This is a different application of the superhet principle, producing an IF that results in an audio-frequency output from the last IF. It is a stable arrangement, and some initial theoretical objections have proved incorrect. It was said, for instance, that it could not handle double-sideband AM without the oscillator producing a signal of identical frequency and also in phase. Amateurs themselves helped to pioneer synchrodyne circuits in which this phase coherence is produced by a technique given the grandiose title of 'exalted carrier detection'. Given a choice, a beginner should choose a good straightforward superhet receiver with plenty of separate waveband circuits, and graduate to the newer designs after gaining experience.

Some devices introduced for quite serious reasons are subsequently found to need even more additional devices to help them overcome operational problems, and this can lead the new amateur into extra expense. For example, while a crystal filter is a good feature, it may be obvious that it results in a high peak of selectivity. Code on CW will be satisfactory, but telephony may be distorted, owing to the narrow band-pass. The 'crystal can be cut', indeed, but a more ambitious type of receiver has two crystal filters operating on slightly different frequencies, which results in a band-pass arrangement (labelled 'Tone Control' on certain US receivers) that does not distort speech channels.

Today a never-ending series of solid-state and semi-conductor devices are transforming receiver circuitry. They are best known by initials. The first to make any impact in the ham world was the FET (field-effect transistor), a device that in fact revolutionised the design of professional television colour cameras. A high degree of amplification in the red, green and blue channels is needed right in the camera head, as close as possible to the Plumbicon or Leddicon tubes, and the introduction of FETs gave this amplification without marked increase in circuit noise. They also have a high input impedance, and are satisfactory low-noise mixers.

The FET has been followed by the IGFET (insulated-gate field-effect transistor), and more recently by a series of MOSTs (metal-oxide semiconductor transistors), including the RCA's MOS-FETs. Some of these solid-state devices are easily harmed by static charges, which can damage the insulation layer, so they are marketed with safety spring clips that in effect short-circuit them until they have been wired in and are ready to operate.

When integrated circuits were first introduced into the computer world, they were virtually too costly for amateur radio. Semiconductor integrated circuits (SICs) are usually formed on a minute slice of silicon, and they incorporate resistors, capacitors and transistors in well-known circuit modules. As mass production and circuit standardisation have drastically cut their cost, they are now coming into amateur radio use. Even hams who build their own equipment are able to make great savings in space, and in the cost of transistors, resistors and capacitors, by using these minute chips with 'all the bits built in'.

Naturally it is not necessary to make use of all the tiny components built into an SIC, and one can use external circuitry for oscillators or other ancillaries that might be affected by heat or voltage instability within the SIC group. Mullard's and Plessey's are among the British manufacturers of SICs suitable for amateur radio, and the amateur journals on both sides of the Atlantic publish receiver circuits incorporating standard SICs produced originally for domestic and car radios. Examples are the SGS TBA651, the RCA CA3088E and the Toshiba TA 7046P.

Output from the receiver is to a matching speaker, or to phones. Of course, when operating his set, the ham usually clamps phones to ears, tunes to his signal, snaps on the speaker switch and waits hopefully for his shack to fill with sound. Inevitably long hours are spent with the phones, so the best is not too good. For the amateur transmitter there are certain advantages in the so-called combined 'mic headset', a miniature boom microphone and an audio headset with a push-to-talk switch. Sets of this type were introduced by companies such as Telex (Sound Research Communications Inc), represented in Canada by Double Diamond Electronics Ltd of Scarborough, Ontario.

In the United Kingdom the name S. G. Brown became synonymous with top quality in amateur and marine headphones in the 1920s when, in the search for the highest sensitivity on faint signals, the company introduced a tuned-reed phone-and-diaphragm arrangement, with a milled knob at the rear of each earpiece to enable the pole-pieces to be set close to the diaphragm. Today S. G. Brown Communications Ltd, with Amplivox, is part of the Racal Electronics group, and among the several headsets suitable for amateur use are the F-type, which has a headband assembly of extremely lightweight anodised duralumin. The total weight of an F-set is $9\frac{1}{2}$oz (267g), and there are types from 15-ohm to 12,000-ohm impedance. Headphones must of course be matched to the output impedance of the receiver.

Outline descriptions of receiver requirements have been given because readers with little technical knowledge will need to know the pros and cons. After beginning to use a communications receiver they will want to be able to modify it (as with a front-end converter) without going to the additional expense of buying a complete new set. It is a different matter with a transmitter, because by the time amateurs will need to select one they will have taken their technical examination (novice class in the USA and possibly the non-code Amateur Sound A or B licences in the United Kingdom), and will have a good idea of what equipment to choose.

Frequency measurement and basic antenna layout have already been covered. The sensible transmitting ham would be wise to start with the MO (master oscillator) end of his equipment, after first deciding which particular facet of transmission interests him most at the outset – CW code, telephony, television, or perhaps single-sideband or frequency-mod transmission.

By international agreement the frequency deviation of an amateur FM station must not be wider than 2·5kHz, bringing it into line with conventional double-sideband AM. Rather a different proposition from some commercial FM broadcasting, it is technically known as 'NBFM' (narrow-band FM). Even though the frequency is varied under modulation, the basic frequency must nevertheless be just as rigidly controlled as in any other form of transmission.

As we have seen, the two essential forms of frequency control in any transmitter are the VFO (variable-frequency oscillator, valve or transistor), and the crystal. The tube- or crystal-controlled master oscillator is really the heart of the ham outfit, and the amateur's devotion to his hobby, and to the true ham spirit, shows itself in the design of his MO. Care taken at this initial stage is reflected all through the transmitter, and by its on-air performance and frequency stability.

Many amateurs begin by building their transmitters, although of course, if money is no object, a wide range of professional transmitters and transceivers is available from the major manufacturers. There are also Heathkits for advanced-design SSB transceivers – a 3W transceiver (on 40 metres) and even a 2m FM transceiver. As detailed instructions are given with each kit, this is a very good way of building up not only the equipment itself but also a sound theoretical and working knowledge of the circuitry.

Although the ordinary crystal is cut to a fixed frequency, so that plug-in crystals are needed to cover a number of different bands, it is possible to use a crystal equivalent of a tubed VFO. This is known as a VXO (variable crystal oscillator), and the effect of slight frequency change ('pulling') is achieved by a series reactance that varies the natural crystal periodicity. With FM transmission, many modern layouts use what is termed a reverse-biased semiconductor diode to apply the varying audio frequency across the controlling crystal. With a tubed MO, a stage known as a reactance tube is linked to the VFO tuned circuit. The audio-frequency voltage is applied to the grid of this valve.

Keying the transmitter is not just a simple on-off process, as beginners might suppose, except that a crystal oscillator can indeed be adequately keyed by cutting the HT supply. Keying a VFO (tubed) may produce undesirable key clicks, so ways are adopted of putting the code key in the grid or screened-grid circuitry of a tubed transmitter.

A good deal of amateur experimentation in CW code transmission has been undertaken on satisfactory keying, and hams have developed a number of what are called 'break-in' systems. This means they can run their receiver and transmitter continuously, the receiver being automatically muted (to avoid heavy overloading) when the key is depressed. Amateurs also design their own transmit-receiver (T-R) antenna switches to shift the aerial from one to another, and this can also be done off the key by suitable relays.

Phone operation is not always, as might be supposed, devoted to producing distortion-free transmission capable of comparison with professional broadcasting. On the contrary, amateur 'fone' can tolerate considerable distortion without losing legibility. If the speech frequencies are cut below about 200 c/s and above about 3,000 c/s, there is a valuable reduction in range of sidebands. Distortion may also be deliberately introduced by 'speech clippers', so that the transmitter can run with better than 100 per cent modulation. There is nothing objectionable about compressing the frequency band, because, as has been shown, the best communication receivers are set to accept only a narrow band on either side of the carrier, and there is no point in transmitting a wide-band carrier that cannot in fact be received and only results in interference with other amateurs on the band.

Grid- and screen-modulation systems are used for modulation, on similar lines to Morse-keying. SSB (single-sideband) transmission has become popular because one half of the band is eliminated and the power in the other half can be increased – theoretically, doubled.

Phone on the HF bands is mostly SSB, which means of course that the results are unintelligible on receivers where the local carrier signal cannot be generated by a BFO, and since all amateur transmission is for experimentation ('part of the self-training of the licensee in communication by wireless', as it says on the British licence) and not for broadcasting, SSB-transmitting amateurs are not unduly dismayed at losing a slice of their listening public.

No matter how expert an amateur may be, there are times when he needs to go back to basic principles. Every shack should have a bookshelf with essential reading matter readily available. This should include the ARRL *Radio Amateur's Handbook*, the RSGB *Amateur Radio Techniques*, and the inexpensive three-volume series of *Radio* by John D. Tucker, AMBritIRE, and Donald Wilkinson, BSc(Eng), AMIEE (English Universities Press), which, despite its brief title, covers the essentials of television as well. In Britain the amateur's bookself would also probably include *Sound and Television Broadcasting* by K. R. Sturley, PhD, BSc, MIEE, which is the BBC engineering training manual, and such practical works of reference as the *Mullard Data Book* and the *Radio Data Reference Book*.

Amateurs who service their own receivers find the RSGB *Amateur Radio Techniques* essential, because its compiler, Pat Hawker (G3VA), has included in the latest editions the only complete list of intermediate frequencies of all British, American, Japanese and ex-military surplus receivers. The IFs are quoted for a wide range of ham receivers, including Collins, Drake, Eddystone, Hallicrafters, Hammarlund, Heathkit, Marconi, National, Plessey, RCA, Sommerkamp, Swan, Tobe, Trio and many more; plus British, Canadian and US surplus equipment, including B36, R1155, BC312 and RAS (modified HRO). This equipment enables owners of older-type communications receivers to line up the IF stages accurately, despite the wide differences between many British, American and Japanese IF standards.

Every keen amateur should join his national society, and perhaps a local radio or TV club. The American ham will seek membership of the ARRL and receive his regular issues of *QST*. Another useful magazine is *Radio Amateur Call Book*, which may be obtained from 608 South Dearborn Street, Chicago 5, Illinois. The British amateur (transmitting, or registered listener) will join the RSGB and get (and file) his monthly copies of *Radio Communication*.

Finally, but most important, there are all those QSL cards to be displayed and filed, visible proof indeed that in the apt words of G3VA, amateur radio is 'a fascinating combination of a scientific hobby, a competitive sport, and an entry into a worldwide fellowship which knows no boundaries of race, class or creed'.

APPENDIX OF MISCELLANEOUS DATA

TECHNICAL TERMS AND ABBREVIATIONS

Term	*Abbreviation*
Alternating current	AC
Ampere	A or amp
Amplification factor	Symbol μ (mu)
Amplitude modulation	AM
Audio frequency	AF
Automatic frequency control	AFC
Automatic gain control	AGC
Bandwidth	Symbol Δf
Beat-frequency oscillator	BFO
Buffer amplifier	BA
Capacitance	Symbol C
Cathode-ray oscilloscope	CRO
Cathode-ray tube	CRT
Centi-	c
Centigrade	C
Continuous wave	CW
Crystal	XTAL
Crystal oscillator	CO
Deci	d
Decibel	dB
Direct current	DC
Direction-finding	DF
Double sideband	DSB

APPENDIX OF MISCELLANEOUS DATA

Electron-coupled oscillator	ECO
Electronic key	ELBUG
Fahrenheit	F
Frequency doubler	FD
Frequency modulation	FM
Gallium arsenide semiconductor	GaAs
Gauss	G
Germanium semiconductor	Ge
Giga-	G
Gramme	g
Ground connection, earth	GND
Henry	H
Hertz	Hz
High frequency	HF
Impedance	Z
Independent sideband	ISB
Inductance-capacitance	L/C
Input	INPT
Intermediate frequency	IF
Kilo-	K
Local oscillator	LO
Long wave	LW
Low frequency	LF
Low tension	LT
Master oscillator	MO
Medium frequency	MF
Mega-	M
Metre	m
Micro	μ
Micro-micro	p
Milli-	m
Modulated continuous wave	MCW
Nano-	n
Narrow-band frequency modulation	NBFM
Ohm	Ω
Peak-to-peak	p-p
Phase modulation	PM
Pico-	p
Power amplifier	PA
Pulse-repetition frequency	PCF
Push-pull	PP
Radio frequency	RF

APPENDIX OF MISCELLANEOUS DATA

Radio telephony	R/T
Reactance	X
Rectified (raw) alternating current	RAC
Receiver	RX
Semi-automatic morse key	BUG
Short wave	SW
Short-wave listener	SWL
Signal frequency	SF
Silicon semiconductor	Si
Single sideband	SSB
Super high frequency	SHF
Television interference	TVI
Transmitter	TX
Travelling wave tube	TWT
Ultra high frequency	UHF
Variable-frequency oscillator	VFO
Very high frequency	VHF
Volts	V
Watts	W
Wavelength	λ
Wireless telegraphy	W/T
Worked all continents	WAC

SEMICONDUCTOR TERMINOLOGY

Enhancement-type field-effect transistor Field-effect transistor that has substantially zero channel conductivity with zero gate to source voltage, and whose channel may be made conductive by the application of a gate to source voltage of appropriate polarity.

Field-effect transistor One in which the current flowing through a conduction channel is controlled by an electric field arising from a voltage applied on a gate electrode.

Metal-oxide semiconductor field-effect transistor (MOST) Insulated-gate field-effect transistor (FET) in which the insulating layers between the gate electrode and channel are made of oxide material.

n-type semiconductor Extrinsic semiconductor in which the conduction-electron density exceeds the mobile hole density.

p-type semiconductor Extrinsic semiconductor in which the mobile hole density exceeds the conduction-electron density.

P/N junction Junction between p- and n-type regions in a semiconductor material.

APPENDIX OF MISCELLANEOUS DATA

Thyristor Silicon-controlled rectifier. Forward conduction only after a signal is applied to the control electrode, and then can be opened only by reduction, removal or reversal of anode supply.

Triode field-effect transistor A field-effect transistor having a gate region, source and a drain region.

Varistor Two-terminal semiconductor device having a non-linear but symmetric current-voltage characteristic.

Zener diode Silicon diode having a very large reverse current above a certain reverse voltage.

TELEVISION STANDARDS

Amateur slow-scan and closed-circuit television sets its own standards, but in every country on-air transmissions tend to adopt the national line and field standards. This table shows characteristics of television systems as indicated in the CCIR Report 308, Xth Plenary Assembly, Geneva.

System	Number of lines	Channel width (MHz)	Vision bandwidth	Vision modulation	Sound modulation
A	405	5	3	Pos	AM
B	625	7	5	Neg	FM
C	625	7	5	Pos	AM
D	625	8	6	Neg	FM
E	819	14	10	Pos	AM
F	819	7	5	Pos	AM
G	625	8	5	Neg	FM
H	625	8	5	Neg	FM
I	625	8	5·5	Neg	FM
K	625	8	6	Neg	FM
K^1	625	4	6	Neg	FM
L	625	8	6	Pos	AM
M	525	6	4·2	Neg	FM
N	625	6	4·2	Neg	FM

UNITED KINGDOM PROFESSIONAL CHANNELS

Carelessly designed and monitored amateur equipment may cause interference with local radio and TV receivers (TVI). For reference, this table gives the frequencies allocated for broadcasting.

APPENDIX OF MISCELLANEOUS DATA

Band	Frequencies	UK usage
LF (long wave)	150–285kHz	Domestic radio. BBC Radio 2
MF (medium wave)	525–1605kHz	Domestic radio. BBC national and local. IBA local. BBC External Services
Band I TV (VHF)	41–68 MHz	Channels 1–5, BBC 1 405-line TV
Band II (VHF)	87·5–100MHz (extended to 104MHz in certain European countries)	VHF radio (including stereo) BBC national and local. IBA local
Band III (VHF)	174–216MHz	Channels 6–13, 405-line TV, BBC 1 and IBA
Band IV (UHF)	470–582MHz	Channels 21–34, 625-line TV, BBC 1, BBC 2 and IBA
Band V (UHF)	614–854MHz (extended to 960MHz outside the UK)	Channels 39–68, 625-line TV BBC 1, BBC 2 and IBA
Band VI (SHF) (super high frequency)	11·7–12·5 GHz	Satellite TV and radio. Terrestrial links

IDENTIFYING RESISTORS

Preferred values of resistors from 10 ohms upwards are as follows:
10, 12, 15, 18, 22, 27, 33, 47, 56, 68, 82 ohms × 1 ... × 10 ... × 100

RESISTOR MARKINGS

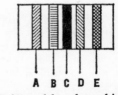

Coloured band marking

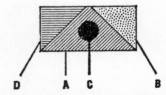

Body, tip and spot marking

APPENDIX OF MISCELLANEOUS DATA

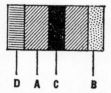

Body, tip and central band marking

etc. These values represent approximately a geometric series increasing by 20 per cent for each value. Resistor values are identified by a standard colour code that takes coloured band, body, tip and spot markings into account.

Colour	1st digit (A)	2nd digit (B)	Multiplier (C)	Percentage tolerance (D)	Grade
Silver			10^{-2}	± 10	
Gold			10^{-1}	± 5	
Black		0	1		
Brown	1	1	10	± 1	
Red	2	2	10^2	± 2	
Orange	3	3	10^3		
Yellow	4	4	10^4		
Green	5	5	10^5		
Blue	6	6	10^6		
Violet	7	7	10^7		
Grey	8	8	10^8		
White	9	9	10^9		
Pink					1
None				± 20	

The first colour A denotes first digit of nominal resistance value, the second colour B denotes second digit of nominal value, the third colour C denotes the multiplier, the fourth colour D denotes the selection tolerance, and the fifth colour E denotes the grade. For example, a resistor with a first band yellow, second band violet, third band red and fourth band silver is 4,700 ohms ± 10 per cent.

APPENDIX OF MISCELLANEOUS DATA

IDENTIFYING CAPACITORS

The value of a capacitor is identified by a standard colour code, but there are five-unit and six-unit codes in use. The five-unit code is as follows:

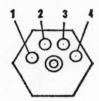

End colour = Temperature coefficient
First colour = Capacitance in pF
Second colour = Significant figures
Third colour = Capacitance multiplier
Fourth Colour = Selection tolerance

Colour	End colour temperature coefficient	First and second colours capacitance (pF) and significant figure	Third colour multiplier	Fourth colour selection tolerance (more than 10 pF)
Black		0	1	± 20 per cent
Brown		1	10	± 1 per cent
Red		2	100	± 2 per cent
Orange		3	1000	± 2·5 per cent
Yellow		4	10 000	
Green		5		± 5 per cent
Blue		6		
Violet	N750 (—750)			
Grey		8	0·01	
White	P100 (+100)	9	0·1	± 10 per cent

APPENDIX OF MISCELLANEOUS DATA

Example: End colour violet -750×10^{-6} per degree C. (1) Red, (2) Violet —270 pF, (3) Brown, (4) Red $\pm$ 2 per cent. If no distinct end colour is marked, or if this is silver, the capacitor has a high permittivity dielectric.

The six-unit colour code for capacitors is as follows:

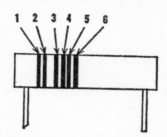

First colour = Temperature coefficient
Fourth colour = Capacitance
Sixth colour = Tolerance

| Colour | *Temperature coefficient* | | *Capacitance (pF)* | | *Selection tolerance* |
	1st colour significant figure	2nd colour multiplier	3rd and 4th colour figure	5th colour multiplier	6th colour (over 10 pF)
Black	0	1	0	1	$\pm$ 20 per cent
Brown	-3	10	1	10	$\pm$ 1 per cent
Red	-8	100	2	100	$\pm$ 2 per cent
Orange	-1.5	1000	3	1000	$\pm$ 2.5 per cent
Yellow	-2.2	10 000	4	10 000	
Green	-3.3				$\pm$ 5 per cent
Blue	-4.7				
Violet	-7.5		7		
Grey	$+3$	0.01	8	0.01	
White	$+1$	0.1	9	0.1	$\pm$ 10 per cent

APPENDIX OF MISCELLANEOUS DATA

INTERNATIONAL WEATHER VOCABULARY

English	Danish	Dutch	French	German	Spanish
Gale	Sturmende Kurling	Storm	Coup de vent	Stürmischer Wind	Viento duro
Storm	Uvejr	Storm	Tempête	Sturm	Temporal
Hurricane	Orkan	Orkaan	Ouragan	Orkan	Huracan
Pressure	Tryk	Druk	Pression	Druck	Presion
Low	Lav	Lagg	Basse	Tief	Baja
High	Høj	Hoog	Haut	Hoch	Alta
Trough	Udlober	Trog	Creux	Troglinie	Vaguada
Ridge	Ryg	Rug	Dorsale	Rücken	Dorsal
Depression	Lavtryk	Minimum	Dépression	Depression	Depresión
Anticyclone	Anticyklon	Gebied van hooge drukkingmaximum	Anticyclone	Hoch	Anticiclón
Cold	Kold	Koud	Froid	Kalt	Frio
Warm	Varm	Warm	Chaud	Warm	Calido
Clouds	Skyer	Wolken	Nuages	Wolken	Nubes
Rain	Regn	Regen	Pluie	Regen	Leuvia
Snow	Sne	Sneeuw	Neige	Schnee	Neive
Drizzle	Støvregn	Motregen	Bruine	Sprühregen	Llovianaa
Hail	Hagl	Hagel	Grêle	Hagel	Granizo
Shower	Byge	Stortbui	Averse	Schauer	Aguacero
Thunder	Torden	Donder	Tonnerre	Donner	Tormenta
Visibility	Synsvidde	Zicht	Visibilité	Sicht	Visibilidad
Fog	Taage	Mist	Brouillard	Nebel	Niebla
Mist	Let taage	Nevel	Brume légère	Nebel	Neblina
Wind	Vind	Wind	Vent	Wind	Viento
Gust	Vindstod	Windstoot	Rafale	Windstoss	Racha
Frost	Frost	Vorst	Gelée	Frost	Escarcha

RST CODE

Code for reporting on quality of Morse signals received. 'R' is Readability, 'S' is Signal strength, and 'T' is tone.

Readability
1. Unreadable
2. Barely readable – odd words distinguishable
3. Readable with considerable difficulty
4. Readable with almost no difficulty
5. Perfectly readable

APPENDIX OF MISCELLANEOUS DATA

Signal strength
1. Faint, hardly audible
2. Very weak
3. Weak signals
4. Fair
5. Fairly good signals
6. Good
7. Moderately strong
8. Strong signals
9. Extremely strong

Morse Tone
1. Rough hissing note
2. Very rough humming tone
3. Rough low-pitched tone
4. Rough hum, some tone
5. Tone and hum equal
6. Good tone, some hum
7. Nearly pure tone
8. Good tone, slight trace hum
9. Pure tone

THE AMATEUR'S Q CODE

The following code can be used on Morse or voice, since it is understood internationally. Code letters can be used both as question and reply.

QAV	Are you calling me? (I am calling . . .)
QRB	How far are you from my stations (miles or km)?
QRG	Will you tell me your exact frequency (kH, MHz)?
QRH	Does my frequency vary?
QRI	How is my tone?
QRJ	Are you receiving me badly? Are signals too weak?
QRK	What is readability of my signals? (See RST code 1–5.)
QRL	Are you busy? (I am busy, do not cut in.)
QRM	Are you being jammed? Interfered with?
QRN	Are you troubled with atmospherics?
QRO	Shall I step up transmitter power?
QRP	Shall I decrease transmitter power?
QRQ	Shall I send faster? (Words per minute . . .)

APPENDIX OF MISCELLANEOUS DATA

QRS	Shall I send slower?
QRT	Shall I stop sending? 'Quiet'. Stop sending.
QRU	Have you anything for me? (I have nothing.)
QRV	Are you ready?
QRW	Shall I tell . . . that you are calling him? (Also 'Please tell . . . that I am calling him'.)
QRX	When will you call again? (. . . hours on . . . MHz.)
QRZ	By whom am I being called?
QSA	What is the strength of my signals?
QSB	Does the strength of my signals vary?
QSD	Is my keying correct? (Also 'Are my signals clear?')
QSG	Shall I send . . . messages at a time?
QSL	Please give acknowledgement of receipt of my signals.
QSM	Shall I repeat the last message I sent?
QSN	Did you hear me on . . . MHz?
QSO	Can you communicate with . . . direct, or through. . . ?
QSP	Will you relay to. . . ?
QSR	Shall I repeat the call on the calling frequency? (Repeat your call on the calling frequency.)
QSS	What working frequency will you use?
QST	This is a general call preceding a message addressed to all amateurs, and specifically to all ARRL members. In effect it is 'CQ ARRL'.
QSU	Shall I reply on this . . . frequency, or on. . . ?
QSV	Shall I transmit a series of VVVVV?
QSW	Will you transmit on this . . . frequency?
QSX	Will you listen for . . . (call-sign) on . . . frequency?
QSY	Shall I change to . . . (frequency) without altering type of transmission?
QSZ	Shall I send each word or group twice?
QTB	Do you agree with my number of words?
QTH	What is your location?
QTR	What is your exact time?
QTU	What hours are you operating?
QTX	Will you keep your station open for further communication, until further notice, or until . . . hours?
QUA	Have you any news of. . . ?

ABBREVIATIONS FOR CODE-WORKING

Introduced into amateur working in the early days of W/T, some of

APPENDIX OF MISCELLANEOUS DATA

these abbreviations are also used colloquially on phone. See also Technical Terms and Abbreviations, p 135.

AA	All after . . . (used generally to request a repetition after a specified point in the message)
AB	All before
ABT	About
ADR	Address
Agn	Again
ANT	Antenna
BCI	Broadcast interference
BCL	Broadcast listener
BK	Break (used to interrupt a transmission in progress)
BN	All between . . .
B4	Before
C	Yes
CFM	Confirm. I confirm
CK	Check
CL	I am closing down
CUD	Could
CUL	See you later
CUM	Come
CW	Continuous-wave transmission
DLVD	Delivered
DX	Distance. Long distance
FB	Fine business. Good working
GA	Go ahead. Resume sending
GB	Good-bye
GBA	Give more detailed address
GD	Good day
GG	Going
GUD	Good
HI	The Morse laugh. Hi!
HR	Hear-hear
HV	Have
HW	How
LID	A poor operator
MSG	Prefix to message
N	No
ND	Nothing doing
NR	Number
NW	Now. I resume sending now

APPENDIX OF MISCELLANEOUS DATA

OB	Old boy
OP	Operator
OT	Old-timer
PBL	Preamble
PSE-PLS	Please
PWR	Power
R	Received solid. OK
RPT	Repeat
SED	Said
SIG	Signal
SKED	Schedule of reception or transmission
TFC	Traffic
TMW	Tomorrow
TNX-TKS	Thanks
UR-URS	Yours
WD-WDS	Words
WUD	Would
WX	Weather
YF (XYL)	Wife
YL	Young lady
73	Best regards
88	Love and kisses

HAM PREFIXES

To make identification possible, International Telecommunications Conferences assign to each nation specific alphabetical blocks, from which stations (not only amateurs) are assigned prefixes. The following list, compiled from official Radio Society of Great Britain sources, is correct at the time of going to press, but political and administrative changes may make subsequent alterations necessary.

A_2	Botswana
A_3	Tonga
A_4	Oman
A_5	Bhutan
A_6	United Arab Emirates
AC_3	Sikkim
AC_4	Tibet
AC_5	Bhutan
AP	Pakistan

APPENDIX OF MISCELLANEOUS DATA

BV	Formosa (Taiwan)
BY	China (and, unofficially, C)
C2	Nauru
C3	Andorra
CE	Chile
CE7Z-CE9	Antarctica (also CE9AA-AM, FB8Y, LA, LU-Z, OR4, UA1, VKØ, VP8, ZL5, ZS1, 8J, 4K)
CEOA	Easter Island
CEOX	San Felix Island
CEOZ	Juan Fernandez Island
CM	Cuba (and CO)
CN8	Morocco
CP	Bolivia
CR3, CR5	(Portuguese) Guinea
CR6	Angola
CR7	Mozambique
CT1	Portugal
CT2	The Azores
CT3	Madeira
CX	Uruguay
DA, DB, DC, DF, DJ, DK, DL, DM	Eastern zone of Germany
DU, DX	The Philippines
EA	Spain
EA6	Balearic Islands
EA8	Canary Islands
EA9	Spanish Morocco
EI	Eire (Southern Ireland)
EL	Liberia
EP, EQ	Persia (Iran)
ET3	Ethiopia
F	France
FC	Corsica
FG	Guadeloupe
FK	New Caledonia
FL	French Somaliland
FM	Martinique
FO	French Oceania (including Tahiti)
FR, FR8	Reunion Island
FU	New Hebrides
FY, FY8	French Guiana
G	England (not the UK generally)
GB	UK, special purposes and exhibitions

148

APPENDIX OF MISCELLANEOUS DATA

GC	Channel Islands (Jersey, Guernsey)
GD	Isle of Man
GI	Northern Ireland
GM	Scotland
GW	Wales
HA, HG	Hungary
HB4, HB9	Switzerland
HE, HBØ	Liechtenstein
HC	Ecuador
HC8	Galapagos Islands
HH	Haiti
HI	Dominica
HK	Colombia
HL, HM	Korea
HP, HO	Panama
HR	Honduras
HS	Thailand
HV	Vatican City, Rome
HZ	Saudi Arabia
I, IA, IB, IC, ID, IE, IF, IG, IH, IL, IP, IZ	Italy
IS, IM	Sardinia
IT	Sicily
JA, JE, JF, JH, JR, KA	Japan
JT	Mongolia
JY	Jordan
K, KN	USA
KC4	US bases in Antarctica
KL7	Alaska
KM6	Midway Island
KP4	Puerto Rico
KS6	American Samoa
KV4	Virgin Islands
KW6	Wake Island
KZ5	Panama Canal zone
LA	Norway
LB	Norway, special purposes and conferences
LU	Argentina
LX	Luxembourg
LZ	Bulgaria
MI	San Marino, Italy
MP4B	Bahrein Island
MP4Q	Qatar

APPENDIX OF MISCELLANEOUS DATA

OA	Peru
OD	Lebanon
OE	Austria
OF, OH	Finland
OHØ	Åland Islands
OK, OM	Czechoslovakia
ON	Belgium
OX	Greenland
OY	Faeroes
OZ	Denmark
PA, PE	The Netherlands (Holland)
PI	Netherlands, special purposes
PJ	Dutch West Indies
PY	Brazil
PZ	Dutch Guiana
S2	Bangladesh
SL	Sweden, special purposes
SM, SK	Sweden
SP	Poland
ST	Sudan
SU	Egypt
SV	Greece, Crete
TA	Turkey
TF	Iceland
TG	Guatemala
TI	Costa Rica
TI9	Cocos Islands
TJ	The Cameroons
TL8	Central African Republic
TN8	Congo Republic (formerly French Congo)
TR8	Gabon Republic
TU2	Ivory Coast
VE, VO	Canada (see detailed list, p 153)
VK	Australia (see detailed list, p 152)
VP7	The Bahamas
VP8	Falkland Islands
VP9	Bermuda
VQ9	Seychelles
VR2	Fiji
VR3	Christmas Island
VR4	Solomon Islands
VR6	Pitcairn Island

APPENDIX OF MISCELLANEOUS DATA

VS5	Brunei
VS6	Hong Kong
VS9o	Sultanate of Oman
VU2	India
XE, XF	Mexico
XU	Cambodia
XW8	Laos
XZ	Burma
YA	Afghanistan
YB, YC, YD	Indonesia
YI	Iraq
YJ	New Hebrides
YK	Syria
YO	Romania
YS	Salvador
YU	Yugoslavia
YV	Venezuela
ZA	Albania
ZB2	Gibraltar
ZD3	Gambia
ZD5	Swaziland
ZD7	St Helena
ZD8	Ascension Island
ZD9	Tristan da Cunha
ZE	Rhodesia
ZF1	Cayman Islands
ZL, ZM	New Zealand
ZL1	Auckland District, NZ
ZL2	Wellington District, NZ
ZL3	Canterbury District, NZ
ZL4	Otago District, NZ
ZL5	New Zealand Antarctica
ZP	Paraguay
ZS	Republic of South Africa
ZS3	South West Africa
3A2	Principality of Monaco (formerly CZ)
3B8	Mauritius
3G3	Chile
3V8	Tunisia (formerly FT4)
4M	Venezuela
4S7	Sri Lanka
4W	Yemen

APPENDIX OF MISCELLANEOUS DATA

4X, 4Z	Israel
5A	Libya
5H1	Zanzibar
5L	Liberia
5N2	Nigeria
5U7	Niger Republic
5V	Togo
5W1	British Samoa
5X5	Uganda
6O	Somali Republic
6Y	Jamaica
7G1	Guinea Republic
7O	Yemen
7Q	Malawi
7X2	Algeria
8F	Indonesia
8P	Barbados
8Q	Maldive Islands
8R	Guyana
9A1	San Marino
9G1	Ghana
9H1	Malta
9J2	Zambia
9K2	Kuwait
9L1	Sierra Leone
9M2, 9M4	West Malaysia
9M6, 9M8	Eastern Malaysia
9N, 9N1	Nepal
9Q5	Zaire
9U5	Burundi
9V1	Singapore
9X5	Rwanda
9Y4	Trinidad and Tobago

Australian Prefixes

VK1-VK8	Australia generally
VK1	Canberra
VK2	New South Wales
VK3	Victoria
VK4	Queensland
VK5	South Australia
VK6	Western Australia

APPENDIX OF MISCELLANEOUS DATA

VK7 Tasmania
VK8 Northern Territories

Canadian Prefixes
VE, VO   Canada generally
VE1   Maritime Provinces
VE2   Quebec Province
VE3   Ontario Province
VE4   Manitoba Province
VE5   Saskatchewan Province
VE6   Alberta Province
VE7   Province of British Columbia
VE8A-L   Yukon Territories
VE8M-Z   North West Territories
VO2   Newfoundland
VO6   Labrador

Soviet Prefixes
RA-RZ   Russian technical stations
UK   Permitted Soviet radio clubs
UK1   Eastern zone (also UA, UK3, UK4, UK6, UV, UW1-6, UN1)
UK2A   White Russia (also UC2, UK2C, UK2I, UK2L, UK2O, UK2S, UK2W)
UK2B   Lithuania (also UP2, UK2P)
UK2F   Kaliningradski (also UA2)
UK2G   Latvia (also UQ2, UK2Q)
UK2R   Estonia (also UR2, UK2T)
UK5O   Moldavia (also UO5)
UK5   Ukraine (also UB5, UT5, UY5)
UK6C   Azerbaijan (also UD6, UK6D, UK6K)
UK6F   Georgia (also UF6, UK6O, UK6Q, UK6V)
UK6G   Armenia (also UG6)
UK7   Kazakhstan (also UL7)
UK8   Uzbekistan (also UI8, UK8A/C/D/F/G/I/L/O/T/U/Z)
UK8H   Turkmenistan (also UH8, UK8E, UK8W, UK8Y)
UK8J   Tadzhikistan (also UJ8, UK8R, UK8S)
UK8M   Kirghizia (also UM8, UK8N, UK8P, UK8Q)
UK9 and Ø Russian Asiatic zone (also UA9 and Ø, UV9 and Ø, and UW9 and Ø)

APPENDIX OF MISCELLANEOUS DATA

US Prefixes

W1	Connecticut, Maine, Massachusetts, New Hampshire, Rhode Island, Vermont
W2	New Jersey, New York
W3	Delaware, Maryland, Pennsylvania and District of Columbia
W4	Alabama, Florida, Georgia, Kentucky, North Carolina, South Carolina, Tennessee, Virginia
W5	Arkansas, Louisiana, Mississippi, New Mexico, Oklahoma, Texas
W6	California
W7	Arizona, Idaho, Montana, Nevada, Oregon, Utah, Washington, Wyoming
W8	Michigan, Ohio, West Virginia
W9	Illinois, Indiana, Wisconsin
WØ	Colorado, Iowa, Kansas, Minnesota, Missouri, Nebraska, North Dakota, South Dakota
KH6	Hawaii
KL7	Alaska

TIME DIFFERENCES

Time differences for major cities, fast or slow, compared with GMT (Greenwich Mean Time). Cities on the longitude of Greenwich keep GMT without conversion.

The following cities are hours *fast* on GMT:

Amsterdam	1	Damascus	2
Ankara	2	Delhi	$5\frac{1}{2}$
Athens	2	Djarkata	$7\frac{1}{2}$
Baghdad	3	Helsinki	2
Bangkok	7	Jerusalem	2
Beirut	2	Karachi	5
Belgrade	1	Kuala Lumpur	$7\frac{1}{2}$
Berne	1	Lagos	1
Bonn	1	Madrid	1
Brussels	1	Manila	8
Bucharest	2	Moscow	3
Budapest	1	Oslo	1
Cairo	2	Paris	1
Canberra	10	Peking	8
Colombo	$5\frac{1}{2}$	Pretoria	2
Copenhagen	1	Rangoon	$6\frac{1}{2}$

APPENDIX OF MISCELLANEOUS DATA

Rome	1	Tokyo		9
Salisbury (Rhodesia)	2	Vienna		1
Stockholm	1	Warsaw		1
Tehran	3½	Wellington (NZ)		12

The following cities are hours *slow* on GMT:

Buenos Aires	3	Quito	5
Havana	5	Rio de Janeiro	3
Mexico City	6	Santiago	4
Montevideo	3	Washington (DC)	5
Ottawa	5		

INTERNATIONAL PHONETIC ALPHABETS

	ICAO (International Civil Aviation Organisation)	American telephone alphabet
A	Alpha	Abel
B	Bravo	Baker
C	Charlie	Charlie
D	Delta	Dog
E	Echo	Easy
F	Foxtrot	Fox
G	Golf	George
H	Hotel	Horn
I	India	Ikon
J	Juliet	Juliet
K	Kilo (Kee-lo)	King
L	Lima (Lee-ma)	Love
M	Mike	Mike
N	November	Nan
O	Oscar	Oboe
P	Papa	Peter
Q	Quebec (Kee-beck)	Queen
R	Romeo	Roger
S	Sierra (See-erra)	Sugar
T	Tango	Tare
U	Uniform	Uncle
V	Victor	Victor
W	Whisky	William
X	X-ray	Eks
Y	Yankee	Yoke
Z	Zulu	Zebra

APPENDIX OF MISCELLANEOUS DATA

Numerical Pronunciation

1	Wun	6	Six
2	Too	7	Sev-en
3	Tree	8	Ait
4	Fo-wer	9	Nine-er
5	Fife	0	Zero

Radio amateurs generally used the modified ICAO code.

USEFUL ADDRESSES

General Amateur Communications
International Amateur Radio Union (Region 1 Division, UK), 51 Pettits Lane, Romford, RM1 4HJ, England.
Radio Society of Great Britain, 35 Doughty Street, London, WC1N 2AE, England.
The American Radio Relay League, Inc, Newington, Conn, USA 06111.
World Organisation of the Scout Movement (Ham Radio, JOTA Organisation), 78 1211 Geneva 4, Switzerland.

Amateur Television
ATA International, Acacialaan 27, B-9720, De Pinte, Belgium.
A-5 Magazine, Box 128, Whitmore Lake, Michigan 48189, USA.
British Amateur Television Club (CQ-TV). Membership: Gordon Sharpley (G6LEE/T), 52 Ullswater Road, Flixton, Urmston, Lancashire. Sales and Library: Grant Dixon (G6AEC/T), Kyrles Cross, Peterstow, Ross-on-Wye, Herefordshire.
CQ Elettronica Magazine, c/o Prof Franco Fanti, Via A. Dallolia 19, 40139 Bologna, Italy.
Ontario ATV Association, 55 Havenbrook Blvd., Willowdale, Ontario, M2JIA7.
TV Amateur (AGAF), 4902 Bad Salzuflen, Pohlmanstr 9, West Germany.
Washington DC (Metrovision) Amateur TV Club, Terry Fox (WB4JFI), PO Box 408, Falls Church, Virginia 22046, U.S.A.

APPENDIX OF MISCELLANEOUS DATA

INTERNATIONAL AMATEUR RADIO UNION

The amateurs who act either as secretaries or liaison officers to their countries' member societies are listed here, and their names, callsigns and addresses are preceded by name of country and society, the latter abbreviated to initials.

Algeria (ARA). Mohamed Benhacine (7X20M), BP No 2 Alger-Care
Austria (OeVSV). Dr F. Stoffel (OE1SFA), A-1014 Vienna, Postbox 999
Belgium (UBA). R. A. Vanmuysen (ON4VY), 52 Diepestraat, 1970 Wezembeck-Oppen, Brabant
Bulgaria (BFRA). T. Chalakov (LZ1BR), PO Box 830, Sofia-C
Cyprus (CARS). R. Whiting (5B4WR), Postbox 1267, Limassol
Czechoslovakia (CRCC). Dr Ondris (OK3EM), Postbox 69, Prague 1
Denmark (EDR). The Secretary, Postbox 79, DK-1003, Copenhagen K
Eire (IRTS). G. Gervin (EI3CC), 6 Montpelier Parade, Blackrock, Co Dublin.
Faeroes (FRA). M. Haasen (OY7ML), Postbox 184, DK-3800, Torshaven
Finland (SRAL). N. R. Kuusisto (OH2XK), Postbox 306, SF-00101 Helsinki, 10
France (REF). C. Laudereau (F9OE), 2 Square Trudaine, 75009 Paris
Ghana (GARS). N. Price (9G1DY), Box 3773, Accra
Great Britain (RSGB). R. F. Stevens (G2BVN), 35 Doughty Street, London WC1 2AE, England
Greece (RAAG). C. Lykiardopoulos (SV1CC), Postbox 564, Athens
Hungary (MRAS). G. Haranyi, 1368 Postbox 214, Budapest 5
Iceland (IRA). I. Thorsteinsson, Postbox 1058, Reykjavik
Israel (IARC). R. Kline (4X4NJ), Gan Yavne 70800
Italy (ARI). S. Pesce (I1ZCT), Via Scarlatti 31, 20124, Milan
Ivory Coast (ARAI). M. Slepoukha (TU2DD), Postbox 20036, Abidjan
Kenya (RSK). The Secretary, Postbox 45681, Nairobi
Lebanon (RAL). E. Eid (OD5FE), Postbox 8888, Beirut
Liberia (LRAA). G. Ruffe (EL2FE), Postbox 1477, Monrovia
Luxembourg (RL). J. Kirsch, 46 rue de la Liberation, Dudelange
Malta (MALTA ARL). C. Warren Falcon (9H1BW), 35 Guardamangia Hill, Guardamangia

APPENDIX OF MISCELLANEOUS DATA

Mauritius (MARS). L. M. Palmyre, Postbox 13, Curepipe
Monaco (ARM). Jean Bardos (3A2EE), 12 rue Bosio, Monte Carlo
Netherlands (VERON). L. van der Nadort (PAØLOU), Loarpak 34, Zundert 4334
Nigeria (NARS). The Secretary, Postbox 2873, Lagos
Norway (NRRL). Jan Gaardsø (LA9IL), Boks 21, Refstad, Oslo 5
Poland (PZK). A. Jeglinski (SP5CM), Postbox 320, Warsaw 1
Portugal (REP). J. H. Gracias (CT1OF), Rue de Parque 46, Lisbon 4
Rhodesia (RSR). Mrs M. Henderson (ZE1JE), Postbox 2377, Salisbury
Romania (FRR). I. Paolazzo (YO3JP), Postbox 1395, Bucharest 5
South Africa (SARL). B. Burger (ZS1IB), Postbox 3911, 6000 Cape Town
Spain (URE). D. José Maria Centeno Perez (EA1HN), Apartado 220, Madrid
Sweden (SSA). Kjell W. Strom (SM6CPI), Mejerigatan 2-232, 2-412 77 Göteburg
Switzerland (USKA). E. Heritier (HB9DX), PO Box 128, CH-4513 Reinach (BL)
USSR (RSF). N. Kazansky (UA3AF), Box 88, Moscow D-362
West Germany (DARC). Amateurfunk-Zentrum des DARC, 3501 Baunatel 1, Postfach 1155
Yugoslavia (SRJ). M. Bogosavljev (YU1SJ), 11000 Belgrade
Zambia (RSZ). B. Clark (9J2CL), PO Box 1596, Lusaka

ACKNOWLEDGEMENTS

In compiling this survey of, and introduction to, amateur radio I have been given the greatest cooperation through the years by many experienced amateurs, and by organisations connected with the ham movement. In particular I am grateful to Pat Hawker, G3VA, an executive of the Royal Television Society and a leading member of the Radio Society of Great Britain, whose own *A Guide to Amateur Radio* and *Amateur Radio Techniques* are classics of their kind for the amateur with electrical knowledge and the ability to read circuit diagrams. We are mutually agreed that there is a vital slot in the amateur-radio publishing world for a book such as this, showing the general reader the wide horizons of this worldwide fellowship.

Great help has been given me by the American Radio Relay League Inc (J. H. Huntoon, W1RW, and Dick Baldwin, W1RU) and by the Staff and Management of Radio Society of Great Britain. I am also indebted to the following individuals and organisations, which are listed alphabetically: Stan Bennett, G3HSC (Belding & Bennett Ltd); Bernard A. Barton, G2HFB; the British Amateur Television Club (Michael J. Sparrow, G6KQJ/T, G8ACB, and Andrew Hughes, Editor *CQ-TV*); the British National Radio & Electronics School (P.T.V. Page, BSc, AMIEE); EMI Electronics Ltd (Colin Woodley and Peter Jones); Douglas A. Findlay, G3BZG; GEC-Marconi Electronics

ACKNOWLEDGEMENTS

and Eddystone Radio Ltd (Peter Baker and Bob Raggett), especially for much information supplied for the Appendix; the Home Office (Radio Regulatory Division); the Independent Broadcasting Authority (F. Howard Steele); the International Amateur Radio Union (Roy F. Stevens, G2BVN, Secretary, Region 1 Division); the Post Office, London; Racal Group Services (W. K. G. Ward and Peter Kennedy); RCA (Government & Commercial Systems, Edward J. Dudley); RCA International Ltd (Leslie Slote); REF, Réseau des Emetteurs Français (P.-L. Trolliet, F5PT, and C. Landereau, F9OE); Henry B. Ruh, WB8HEE (*A-5 Amateur Television Magazine*); UBA, Union Belge des Amateurs-Emetteurs; and the World Organisation of the Scout Movement (L. F. Jarrett, JOTA Organiser, HB9AMS).

The illustrations have kindly been supplied by the following:
American Radio Relay League Inc: plates 1, 3, 5, 10, 16
Amplivox: plate 15
British Amateur Television Club: plate 14
GEC-Marconi Electronics: plates 4 and 9
International Amateur Radio Union: plate 11
Radio Society of Great Britain: plates 2, 6, 7, 8 and 12
RCA International Ltd: plate 17

K. U.

INDEX

Abbreviations, technical terms, 135–7
Activities of radio hams, 26–34, 37–47
Addresses, 156–8
Advanced licence (USA), 23, 25, 98
Age limit for licence, 24–5, 80
Amateur licences, 22–5, 32, 60, 78–99, 108
Amateur Radio Certificate, 94–5
Amateur Radio Public Service, 39
Amateur Radio Technique, 133, 134
Amateur satellite service, 28, 30, 33–4
Amateur Television Association, 64–5
Amateur Television Licence, 80, 81
American Radio Relay League, 10, 23–5, 34, 37, 42
Amplitude modulation, 43–7, 85
Antennas, 111, 115–20
Armstrong, Major Edwin H., 26–7
Audio limits, human, 44
Australian code prefixes, 152–3

Bailey, George W., 28–9
Baird, John Logie, 50, 52
Blind operators, 14, 70
Books and journals, 12, 14, 24–5, 32–3, 57–9, 60–2, 64–6, 89–90, 97, 99–100, 102–3, 106, 112, 117–18, 124, 127, 133–4
British Amateur Television Club, 60–3; equipment at cost price, 61–2
British Broadcasting Corporation club, 39
British Receiving Station number, 91

Canadian code prefixes, 153
Capacitor identification, 141–2
Chess playing, 14
City and Guilds examinations, 92–5, 108
Code-working abbreviations, 145–7
'Conditional' licence (USA), 23, 98, 108
Continuous wave, 43–7

De Forest, Dr Lee, 27
Deloy, Leon, 10, 11
Didah language, 100–9

E-layers, 112
EMI Ltd club, 39
Examinations for licences, 23, 24–5, 60, 80–1, 91, 91–9
Extra Class licence (USA), 23, 99

INDEX

Fast-scan television, 64
Fees, examination and licence, 25, 60, 80, 92
F-layers, 112
Flying-spot scanner, 61, 62
'Four-sevens' patent, 19
Frequency, definition of wave, 20–1; *see also* Wavebands
Frequency modulation (FM), 43–7, 85
Frequency-standard transmissions, 89–90
Fultograph, 41, 43

General ('Conditional') licence (USA), 23, 98, 108
Girl Guide, Japanese amateur, 69

Headphones, 130–1
Heathkit outfits, 123–4, 132
Hertz, Heinrich, 16, 19, 21
Hertz measurements, 21

Integrated circuits, 129–30
Interference, electrical, 43–4, 125
International Amateur Radio Union, 29ff; addresses, 33, 41, 156
International Geophysical Year (1957–8), 28
International Quiet Sun Years (1958–9), 28
International Telecommunications Union, 29–30
International weather vocabulary, 143
Ionosphere, 110–12; prediction maps, 111–12

Jamboree-on-the-Air, 66–70, 73–7
Jenkins, C. Francis, 48–9

Lake Success station, 28–9
Length, definition of wave, 20
Licence regulations, 22–5, 60, 78–99, 108; South Africa, 32

Light waves, 20
Lodge, Sir Oliver, 15–16, 19
Log books, 83
Long wave, definition, 20

Marconi company club, 13
Marconi, Guglielmo, 10, 15–16, 19
Metric measurement of waves, 20–1
Military Affiliate Radio System (MARS), 34, 37
Mitchell, Leslie, 67–9, 70, 73
Morse code, 44, 84, 100–9; in licence tests, 23, 25, 60, 80–1, 94–5, 96, 97
Morse tone code, 144
Mullard Data Book, 134
Mullard Group club, 39

National alphabetical code prefixes, 147–54
Nipkow, Dr Paul, 48
Novice licence (USA), 23, 95–6, 108
Numerical pronunciation, 156

OSCARS, 9, 13, 14, 28, 30, 32, 33–4, 51
Overseas Receiving Station number, 91

Penalties, 79
Permits, *see* Licence regulations
Phonetic alphabet, 155; numerical pronunciation, 156
Photoradiogram (1924), 50
Physically handicapped operators, 14
Pirate radio stations, 79, 82
Plessey Company Ltd club, 39
Post Office club, 39
Professional channels (UK), 138–9
Pulse modulation, 85

Q code, 103, 105, 144–5
Quartz crystal oscillator, 87–91

Racal Group (Services) club, 39

INDEX

Radio, 133
Radio Amateur Call Book, 134
Radio Amateur Examination, 91–5, 108
Radio Amateur Satellite Corporation, 30, 33–4
Radio Amateur's Handbook, 133
Radio Communication, 134
Radio Date Reference Book, 134
Radio Society of Great Britain, 11, 14, 34, 91
Radio waves, 'speed' of, 20
RCA company club, 13, 39–40; and early television, 49–50
Reaction, in signals, 26–7
Readability code, 143
Receiver construction, 122–31; Heathkits, 123
Regeneration, in signals, 26–7
Reinhartz, John, L., 10
Resistor markings, 139–40
Resistor values, 139–40
Righi, Professor Auguste, 16
RST code, 143–4

Satellites, 9, 13, 14, 28, 30, 32, 33–4, 39–40, 139
Scouts (boy) as radio amateurs, 66–70, 73–7
Semiconductor terminology, 137–8
Short wave, definition of, 20
Signal strength code, 144
Simmonds, E. J., 11, 12
Single sideband transmission (SSB), 43, 46, 56
Slow-scan television, 51, 56, 57, 60, 61, 64, 65
(Sound) licence, 80, 81, 86, 87
(Sound mobile) licence, 80, 81
Sound and Television Broadcasting, 133
Soviet code prefixes, 153
'Spark' transmission, 43–4
'Speed' of radio and light waves, 20
Super high frequency bands, 21, 139

Sync (synchronism), 50, 55–6

Technical terms, abbreviations, 135–7
Technician class licence (USA), 23, 96, 108
Television, 48–52, 55–65; frequencies, 50–51, 58, 59; professional channels, 60; technicalities, 51–2, 55–8
Television licence, 80, 81
Television standards, 138
Television weather reception, 39–40
Time differences, 154–5
Transmitters, 85, 131–3; Heathkits, 132
Tuning, 15, 19ff, 122, 126

United Nations Amateur Radio Club, 28–9
USA code prefixes, 154
USA examinations, 23, 25, 95–9, 108

Variable capacitors, 15
Variable condensers, 15
Velocity of propagation, 20

Wavebands, radio: amateur, 22–5, 28, 80–1, 86–7, 113–14; professional, 138–9
Wavebands, television: amateur, 58, 59; professional, 60, 138–9
Wave frequency, definition, 20
Wave length, definition, 20
Wavemeters, 88–9
Wave 'speed', 20
Weather satellites, amateur reception of, 39–40
Weather vocabulary, international, 143
World Administrative Radio Conference for Space Telecommunications, 28